CW00403986

The Story of the Solar System

By

George F. Chambers

Published by Forgotten Books 2012
Originally Published 1904

PIBN 1000071292

THE STORY OF THE SOLAR SYSTEM

BY

GEORGE F. CHAMBERS, F.R.A.S.

OF THE INNER TEMPLE, BARRISTER-AT-LAW

AUTHOR OF THE STORY OF THE STARS

WITH TWENTY-EIGHT ILLUSTRATIONS

NEW YORK

McCLURE, PHILLIPS & CO.

MCMIV

PREFACE.

HAVING in my "Story of the Stars" told of far distant suns, many of them probably with planets revolving around them, I have in the present volume, which is a companion to the former one, to treat of the Sun in particular—our Sun as we may call him—and the body of attendants which own his sway by revolving round him. The attendants are the planets, commonly so called, together with a certain number of comets. I shall deal with all these objects rather from a descriptive and practical than from a speculative or essay point of view, and with special reference to the convenience and opportunities of persons possessing, or having access to, what may be called popular telescopes—telescopes say of from two to four inches of aperture, and costing any sum between £10 and £50. There is much pleasure and profit to be got out of telescopes of this type, always presuming that they are used by persons possessed of patience and perseverance. It is a very great mistake, though an extremely common one, to suppose that unless a man can command a big telescope he can do no useful work, and derive no pleasure from his work. To all such croakers I always point as a moral the achievements of Hermann Goldschmidt, who from an attic window at Fontenay-aux-Roses near Paris, with a telescope of only 2½ inches aperture, discovered no fewer than 14 minor planets.

As this volume is intended for general reading, rather than for educational or technical purposes, I have kept statistical details and numerical expressions within very narrow limits, mere figures being always more or less unattractive.

John Richard Green, in the Preface to his book on *The Making of England*, writes as follows:—"I may add, in explanation of the reappearance of a few passages . . . which my readers may have seen before, that where I had little or nothing to add or to change, I have preferred to insert a passage from previous work, with the requisite connections and references, to the affectation of rewriting such a passage for the mere sake of giving it an air of novelty." I will venture to adopt this thought as my own, and to apply it to the repetition, here and there, of ideas and phrases which are already to be found in my *Handbook of Astronomy*.

G. F. C.

NORTHFIELD GRANGE,
EASTBOURNE, 1895.

CONTENTS.

LIST OF ILLUSTRATIONS.

THE STORY OF THE SOLAR SYSTEM.

CHAPTER I.

INTRODUCTORY STATEMENT.

By the term "Solar System" it is to be understood that an Astronomer, speaking from the standpoint of an inhabitant of the Earth, wishes to refer to that object, the Sun, which is to him the material and visible centre of life and heat and control, and also to those bodies dependent on the Sun which circulate round it at various distances, deriving their light and heat from the Sun, and known as planets and comets. The statement just made may be regarded as a general truth, but as the strictest accuracy on scientific matters is of the utmost importance, a trivial reservation must perhaps be put upon the foregoing broad assertion. There is some reason for thinking that possibly one of the planets (Jupiter) possesses a little inherent light of its own which is not borrowed from the Sun; whilst of the comets it must certainly be said that, as a rule, they shine with intrinsic, not borrowed light. Respecting these reservations more hereafter.

The planets are divided into "primary" and "secondary." By a "primary" planet we mean one which directly circulates round the Sun; by a "secondary" planet we mean one which in the first instance circulates round a primary planet, and therefore only in a secondary sense circulates round the Sun. The planets are also "major" or "minor"; this, however, is only a distinction of size.

The secondary planets are usually termed "satellites," or, very often, in popular language, "moons," because they own allegiance to their respective primaries just as our Moon—*the* Moon —does to the Earth. But the use of the term "moon" is inconvenient, and it is better to stick to "satellite."

There is yet another method of classifying the planets which has its advantages. They are sometimes divided into "inferior" and "superior." The "inferior" planets are those which travel round the Sun in orbits which are inside the Earth s orbit; the "superior" planets are those whose orbits are outside the Earth.

The following is an enumeration of the major planets in the order of their distances, reckoning from the Sun outwards:—

1. Mercury.
2. Venus.
3. The Earth.
4. Mars.
5. Jupiter.
6. Saturn.
7. Uranus.
8. Neptune.

All the above are major planets and also primary planets. In between Nos. 4 and 5 circulate the "Minor" planets, an ever-increasing body, now more than 400 in number, but all, except one or perhaps two, invisible to the naked eye.

The "Inferior" planets it will be seen from the above table comprise Mercury and Venus, whilst the "Superior" planets are Mars and all those beyond.

Great differences exist in the inclinations of the orbits of the different planets to the plane of the ecliptic, a fact which is better shown by a diagram than by a table of mere figures. The orbit of Uranus is indeed so much inclined that its motion is really *retrograde* compared with the general run of the planets: and the same remark applies, though much more forcibly, to the case of Neptune.

The actual movements of the planets round the Sun are extremely

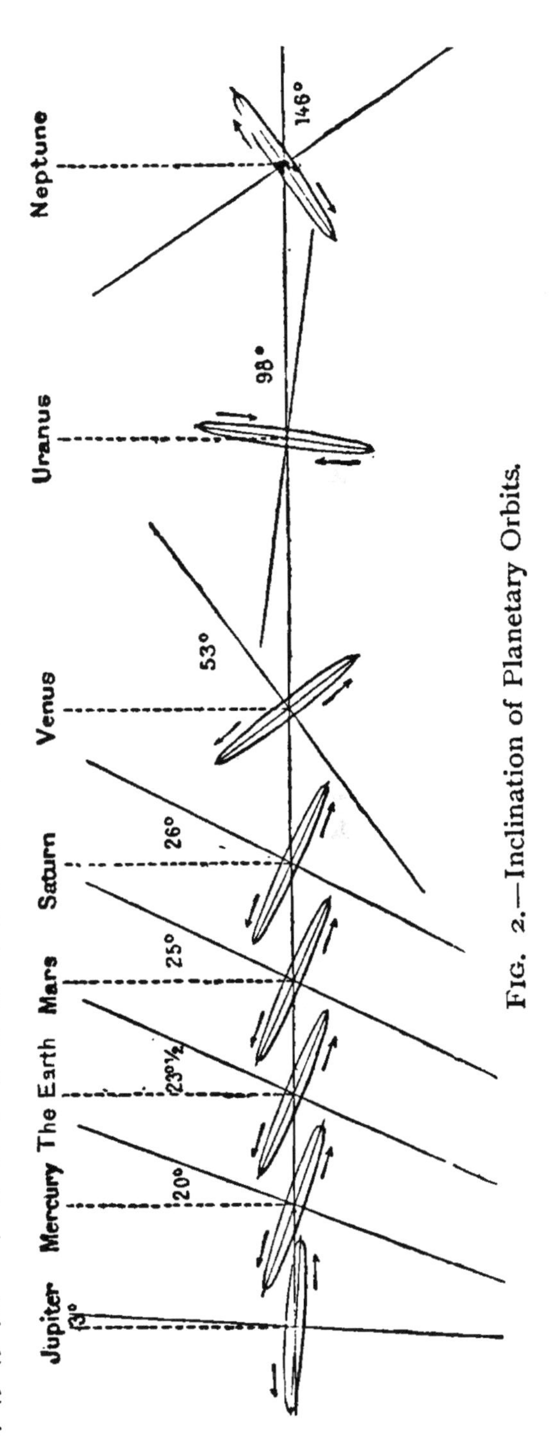

FIG. 2.—Inclination of Planetary Orbits.

simple, for they do nought else but go on, and on, and on, incessantly, always in the same direction, and almost, though not quite, at a uniform pace, though in orbits very variously inclined to the plane of the ecliptic. But an element of extreme complication is introduced into their apparent movements by reason of the fact that we are obliged to study the planets from one of their own number, which is itself always in motion.

If the Earth itself were a fixture, the study of the movements of the planets would be a comparatively easy matter, whilst to an observer on the Sun it would be a supremely easy matter.

Greatly as the planets differ among themselves in their sizes, distances from the Sun, and physical peculiarities, they have certain things in common, and it will be well to make this matter clear before we go into more recondite topics. For instance, not only do they move incessantly round the Sun in the same direction at a nearly uniform pace, but the planes of their orbits are very little inclined to the common plane of reference, the ecliptic, or to one another.* The direction of motion of the planets as viewed from the north side of the ecliptic is contrary to the motion of the hands of a watch. Their orbits, unlike the orbits of comets, are nearly circular, that is, they are only very slightly oval. Agreeably to the principles of what is known as the Law of Universal Gravitation, the speed with which they move in their orbits is greatest in those parts

---

* The remark in the text applies to all the major planets and to a large number of the minor planets, but certain of the minor planets travel in orbits which are considerably inclined to the ecliptic, and therefore to all the other planets.

which lie nearest the Sun, and least in those parts which are most remote from the Sun; in other

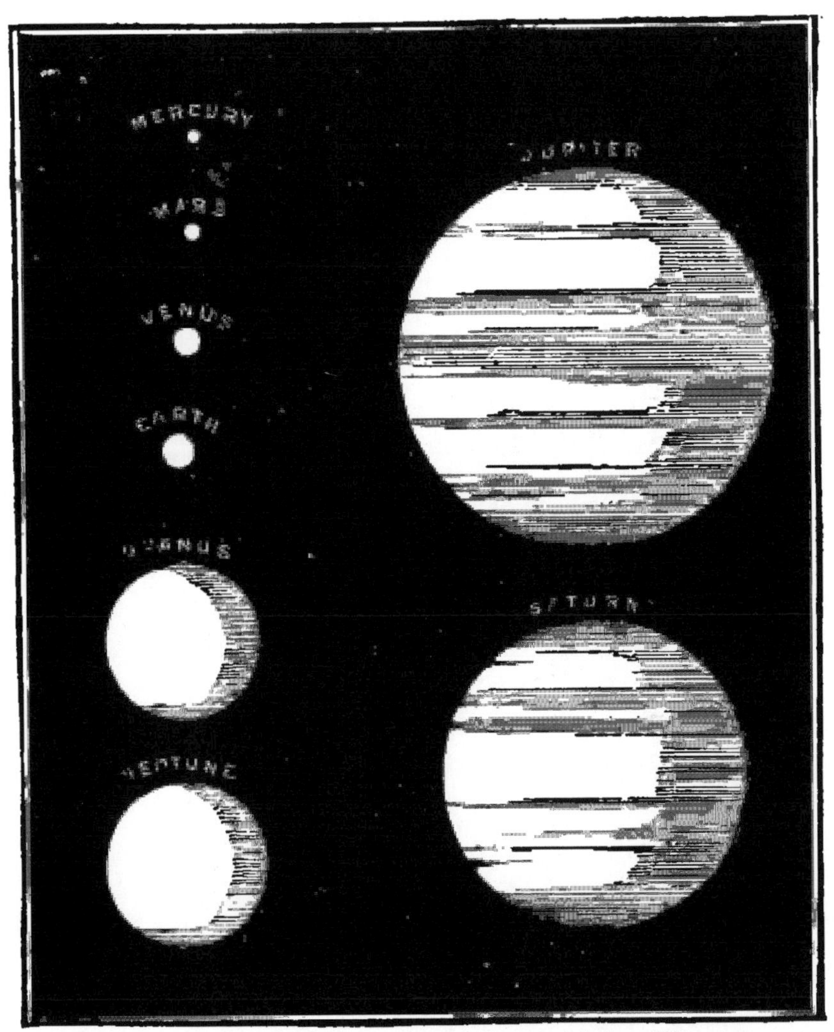

FIG. 3.—Comparative Sizes of the Major Planets.

words, they move quickest in Perihelion and slowest in Aphelion.

The physical peculiarities which the planets have in common include the following points:—they are opaque bodies, and shine by reflecting light which they receive from the Sun. Probably all of them are endued with an axial rotation, hence their inhabitants, if there are any, have the alternation of day and night, like the inhabitants of the Earth, but the duration of their days, measured in absolute terrestrial hours, will in most cases differ materially from the days and nights with which we are familiar.

I stated on a previous page that, owing to the circumstances in which we find ourselves on the Earth, the apparent and real movements of the planets are widely different. It would be beyond the scope of this little work to go into these differences in any considerable detail; suffice it then to indicate only a few general points. In the first place, an important distinction exists between the visible movements of the inferior and superior planets. The inferior planets, Mercury and Venus, lying as they do within the orbit of the Earth, are much restricted in their movements in the sky. We can never see them except when they are more or less near to the rising (or risen) or setting (or set) Sun. The extreme angular distance from the Sun in the sky to which Mercury can attain is but 27°, and therefore we can never observe it otherwise than in sunlight or twilight, for it never rises more than 1½ hours before sunrise nor sets later than 1½ hours after sunset. Of course between these limits the planet is above the horizon all the time that the Sun is above the horizon, but except in very large telescopes is not usually to be detected during the day-time. These remarks regarding Mercury apply likewise in principle to Venus; only the orbit of Venus being larger than the orbit of Mercury, and Venus itself being larger in size than Mercury, the application of these principles leads to somewhat different results. The greatest possible distance of Venus may be 47° instead of Mercury's 27°. Venus is therefore somewhat more emancipated from the effects of twilight. The body of Venus being also very much larger and brighter than the body of Mercury, it may be more often and more easily detected in broad daylight.

It follows from the foregoing statement that the inferior planets can never be seen in those regions of the heavens which are, as it is technically called, in "Opposition" to the Sun; that is, which are on the meridian at midnight whilst the Sun is on the meridian in its midday splendour to places on the opposite side of the Earth. On the other hand, the two inferior planets on stated, though rare, occasions exhibit to a terrestrial spectator certain phenomena of great interest and importance in which no superior planet can ever take part. I am here referring to the "Transits" of Mercury and Venus across the Sun. If these planets and the Earth all revolved round the Sun exactly in the plane of the ecliptic, transits of these planets would be perpetually recurring after even intervals of only a few months; but the fact that the orbit of Mercury is inclined 7°, and that of Venus about 3½, to the ecliptic, involves such complications that transits of Mercury only occur at unequal intervals of several years, whilst, in extreme cases, more than a century may elapse between two successive transits of Venus. For a transit of an inferior planet over the Sun to take place, the Earth and the planet and the Sun must be exactly in the same straight line, reckoned both vertically and horizontally. Twice in every revolution round the Sun an inferior planet is vertically in the same straight line with the Earth and the Sun; and it is said to be in "inferior conjunction" when the planet comes between the Earth and the Sun; and in "superior conjunction" when the planet is on the further side of the Sun, the Sun intervening between the Earth and the planet. But for all three to be horizontally in the same straight line is quite another matter. It is the orbital inclinations of Mercury and Venus which enable them, so to speak, to dodge an observer who is on the lookout to see them pass exactly in front of the Sun, or to disappear behind the Sun; and so it comes about that a favourable combination of circumstances which is rare is needed before either of the aforesaid planets can be seen as round black spots passing in front of the Sun. A passage of either of these planets behind the Sun could never be seen by human eye, because of the overpowering brilliancy of the Sun's rays, even though an Astronomer might know by his calculations the exact moment that the planet was going to pass behind the Sun.

When an inferior planet attains its greatest angular distance from the Sun, as we see it (which I have already stated to be about 27° in the case of Mercury and 47° in the case of Venus), such planet is said to be at its "greatest elongation," "east" or "west," as the case may be. At eastern elongation or indeed whenever the planet is east of the Sun, it is, to use a familiar phrase, an "evening star"; on the other hand, at western elongation, or whenever it is on the western side of the Sun, it is known as a "morning star."

If the movements of an inferior planet are followed sufficiently long by the aid of a star map, it will be seen that sometimes it appears to be proceeding in a forward direction through the signs of the zodiac; then for a while it will seem to stand still; then at another time it will apparently go backwards, or possess a retrograde motion. All these peculiarities have their originating cause in the motion of the Earth itself, for the absolute movement of the planet never varies, being always in the same direction, that is, forwards in the order of the signs.

Turning now to the superior planets, we have to face an altogether different succession of circumstances. A superior planet is not, as it were, chained to the Sun so as to be unable to escape beyond the limits of morning or evening twilight; it may have any angular distance from the Sun up to 180°, reaching which point it approaches the Sun on the opposite side, step by step, until it again comes into conjunction with the Sun. As applied to a superior planet, the term "conjunction" means the absolute moment when the Earth and the Sun and the planet are in the same straight line, the Sun being in the middle. In such a case, to us on the Earth the planet is lost in the Sun's rays, whilst to a spectator on the planet the Earth would appear similarly lost in the Sun's rays, as the Earth would be at that stage of her orbit which we, speaking of *our* inferior planets, call superior conjunction.

For a clear comprehension of all the various matters which we have just been speaking of, a careful study of diagrams of a geometrical character, or better still, of models, would be necessary.

Something must now be said about the phases of the planets. Mercury and Venus, in regard to their orbital motions, stand very much on the same footing with respect to the inhabitants of the Earth as the Moon does, and accordingly both those planets in their periodical circuits round the Sun exhibit the same succession of phases as the Moon does. In the case, however, of the superior planets things are otherwise. Two only of them, Mars and Jupiter, are sufficiently near the Earth to exhibit any phase at all. When they are in quadrature (*i. e.*, 90° from the Sun on either side) there is a slight loss of light to be noticed along one limb. In other words, the disc of each ceases for a short time, and to a slight extent, to be truly circular; it becomes what is known as "gibbous." This occasional feature of Mars may be fairly conspicuous, or, at least, noticeable; but in the case of Jupiter it will be less obvious unless a telescope of some size is employed.

If the major planets are arbitrarily ranged in two groups, Mercury, Venus, the Earth and Mars being taken as an interior group, comparatively near the Sun; whilst Jupiter, Saturn, Uranus and Neptune are regarded as an exterior group, being at a great distance from the Sun, it will be found that some important physical differences exist between the two groups.

Of the interior planets, the Earth and Mars alone have satellites, and between them make up a total of only three. The exterior planets, on the other hand, all have satellites, the total number being certainly seventeen, and possibly eighteen. In detail, Jupiter has four, Saturn eight, Uranus four, and Neptune one, and perhaps two. These facts may be regarded as an instance of the beneficence of the Creator of the Universe if we consider that the satellites of these remoter planets are so numerous, in order that by their numbers they may do something to make up for the small amount of light which, owing to their distance from the Sun, their primaries receive. Then again, the average density of the first group

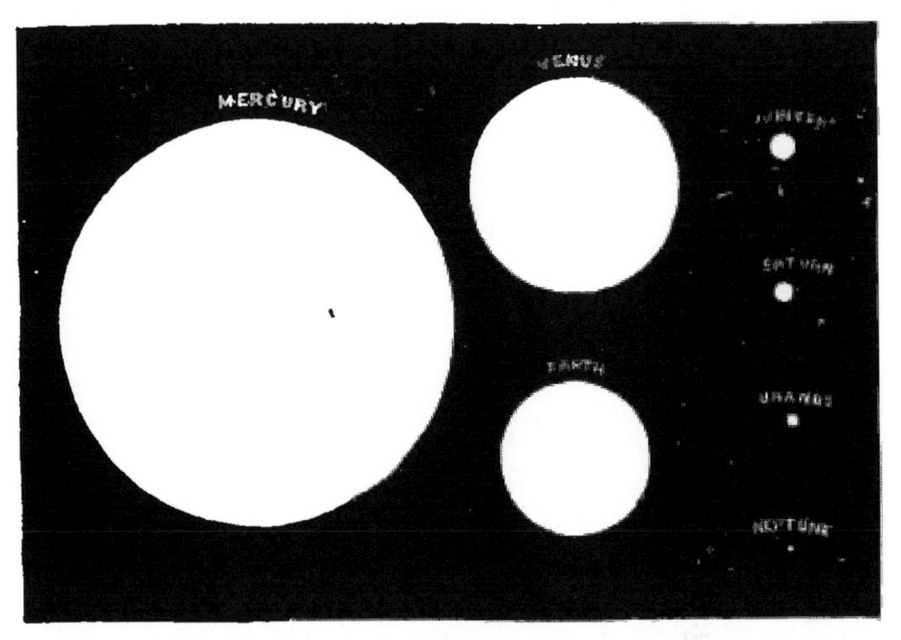

FIG. 4.—Comparative size of the Sun as seen from the Planets named.

of planets greatly exceeds the average density of the second group in the approximate ratio of 5 to 1. Finally, there is reason to believe that a marked difference exists in the axial rotations of the planets forming the two groups. We do not know the precise figures for all the exterior planets, but the knowledge which we do possess seems to imply that the average length of the day in the case of the interior planets is about twenty-four hours, but that in the case of the exterior planets it is no more than about ten hours. These figures can, however, only be presented as possibly true, because observations on the rotation periods of Mercury and Venus on the one hand, and of Uranus and Neptune on the other, are attended with so much difficulty that the recorded results are of doubtful trustworthiness. It is, however, reasonable to presume that the actual size of the respective planets has more to do with the matter than their distances from the Sun.

I think that the foregoing summary respecting the planets collectively embraces as many points as are likely to be of interest to the generality of readers; we will therefore pass on to consider somewhat in detail the several constituent members of the solar system, beginning with the Sun.

CHAPTER II.

THE SUN.

THERE was once a book published, the title of which was "The Sun, Ruler, Fire, Light and Life of the Planetary System." The title was by no means a bad one, for without doubt the Sun may fairly be said to represent practically all the ideas conveyed by the designations quoted.

There is certainly no one body in creation which is so emphatically pre-eminent as the Sun. Whether or no there are stars which are suns—centres of systems serving in their degree the purposes served by our Sun, I need not now pause to enquire, though I think the idea is a very probable one; but of those celestial objects with which our Earth has a direct relationship, beyond doubt the Sun is unquestionably entitled to the foremost place. It is, as it were, the pivot on which the Earth and all the various bodies comprising the Solar System revolve in their annual progress. It is our source of light and heat, and therefore may be called the great agent by which an Almighty Providence wills to sustain animal and vegetable life. The consideration of all the complicated questions which arise out of these functions of the Sun belongs to the domain of Physics rather than that of Astronomy; still these matters are of such momentous interest that an allusion to them must be made, for they ought not to be lost sight of by the student of Astronomy. Half a century ago the actual state of our knowledge respecting the Sun might without difficulty be brought within the compass of a single chapter in any book on Astronomy, but so enormous has been the development of knowledge respecting the Sun of late years, that it is no longer a question of getting the materials properly into one chapter, but it is a matter of a whole volume being devoted to the Sun, or even, as in the case of Secchi, of two large octavo volumes of 500 pages each being required to cover the whole ground exhaustively. The reader will therefore easily understand that in the space at my disposal in this little work nothing but a passing glimpse can possibly be obtained of this great subject. It is great not only in regard to the vast array of purely astronomical facts which are at a writer's command, but also on account of the extensive ramifications which the subject has into the domains of chemistry, photography, optics and cognate sciences. I shall therefore endeavour to limit myself generally to what an amateur can see for himself with a small telescope, and can readily understand, rather than attempt to say a little something about everything, and fail in the effort.

The mean distance of the Earth from the Sun may be taken to be about 93 millions of miles, and this distance is employed by Astronomers as the unit by which most other long celestial distances are reckoned. The true diameter of the Sun is about 866,000 miles. The surface area exceeds that of the Earth 11,946 times, and the volume is 1,305,000 times greater. The mass or weight of the Sun is 332,000 times that of the Earth, or about 700 times that of all the planets put together. Bulk for bulk the Sun is much lighter than the Earth: whilst a cubic foot of the Earth on an average weighs rather more than 5 times as much as a cubic foot of water, a cubic foot of Sun is only about 3½ times the weight of the same bulk of water. This consideration of the comparative lightness of the Sun (though in his day the Sun was thought to be lighter than it is now supposed to be) led Sir J. Herschel to infer that an intense heat prevails in its interior, independent it may be of its surface heat, so to speak, of which alone we are directly cognizant by the evidence of our senses.

The Sun is a sphere, and is surrounded by an extensive but attenuated envelope, or rather series of envelopes, which taken together bear some analogy to the atmosphere surrounding the Earth. These envelopes, which we shall have to consider more in detail presently, throw out rays of light and heat to the confines of the Solar System, though as to the conditions and circumstances under which that light and heat are generated we are entirely ignorant. Of the potency of the Sun's rays we can form but a feeble conception, for the amount received by the Earth is, it has been calculated, but one 2300-millionth of the whole. Our annual share would, it is supposed, be sufficient to melt a layer of ice spread uniformly over the Earth to a depth of 100 feet, or to heat an ocean of fresh water 60 feet deep from freezing point to boiling point. The illuminating power of the Sun has to be expressed in language of similar profundity. Thus it has been calculated to equal that which would be afforded by 5563 wax candles concentrated at a distance of one foot from the observer. Again, it has been concluded that no fewer than half a million of full moons shining all at once would be required to make up a mass of light equal to that of the Sun. I present all these conclusions to the reader as they are furnished by various physicists who have investigated such matters, but it is rather uncertain as to how much reliance can safely be placed on such calculations in detail.

To an amateur possessed of a small telescope, the Sun offers (when the weather is above the English average of recent years) a very great and constant variety of matters for studious scrutiny in its so-called "spots." To the naked eye, or even on a hasty telescopic glance, the Sun presents the appearance of a uniform disc of yellowish white colour, though often a little attention will soon result in the discovery of a few, or it may be many, little black, or blackish patches, scattered here and there over the disc seemingly without order or method. We shall presently find out, however, that this last-named suggestion is wholly inaccurate. Though commonly called "spots," these dark appearances are not simple spots, as the word might imply, for around the rather black patch which constitutes generally the main feature of the spot there is almost invariably a fringe of paler tint; whilst within the confines of the black patch which first catches the eye there is often a nucleus or inner portion of far more intense depth of shade. The innermost and

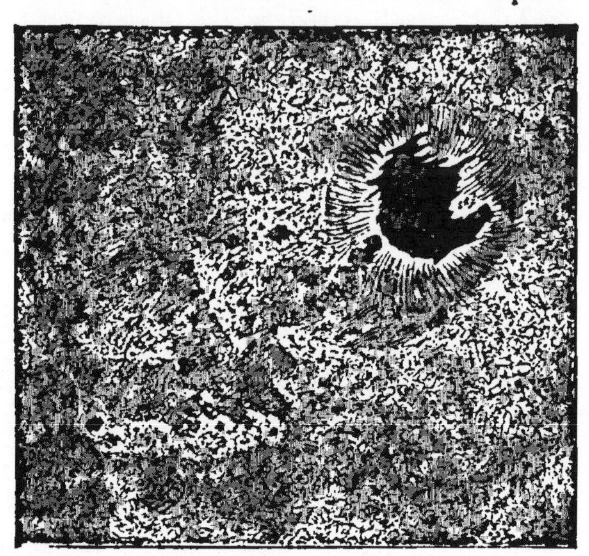

FIG. 5.—Ordinary Sun-spot, June 22, 1885.

darkest portion being termed the *nucleus*, the ordinary black portion is known as *umbra*, whilst the encompassing fringe is the *penumbra*. It is not always the case that each individual umbra has a penumbra all to itself, for several spots are occasionally included in one common penumbra. And it may further be remarked that cases of an umbra without a penumbra and the contrary are on record, though these may be termed exceptional, often having relation to material organic changes either just commencing or just coming to a conclusion. A marked contrast subsists in all cases between the luminosity of the penumbra and that of the general surface of the Sun contiguous. Towards its exterior edge the penumbra is usually darker than at its inner edge, where it comes in contact with the umbra. The outline of the penumbra is usually very irregular, but the umbra, especially in the larger spots, is often of regular form (comparatively speaking of course) and the nucleus (or nuclei) of the umbra still more noticeably partakes of a compactness of outline.

Spots are for the most part confined to a zone extending 35° or so on each side of the solar equator; and they are neither permanent in their form nor stationary in their position. In their want of permanence, they are subject, apparently, to no definite laws, for they frequently appear and disappear with great suddenness.

Their motions are evidently of a two-fold nature; the Sun itself rotates on its axis, and the spots collectively participate in this movement of rotation; but over and above this it has been conclusively proved that sometimes a spot has a proper motion of translation of its own independently of the motion which it has in consequence of the Sun's axial rotation. Curiously enough, spots are very rare immediately under the Sun's equator. It is in the zone extending from 8° to 20° North or South, as the case may be, that they are most abundant; or, to be more precise still, their favourite latitude seems to be 17° or 18°. They are often more numerous and of a greater general size in the northern hemisphere, to which it may be added that the zone between 11° and 15° North is particularly noted for large and enduring spots. A gregarious tendency is often very obvious, and where the groups are very straggling an imaginary line joining the extreme ends of the group will generally be found more or less parallel to the solar equator; and not only so, but extending a long way, or sometimes almost entirely, across the whole of the visible disc.

With respect to the foregoing matters Sir John Herschel remarked :—" These circumstances . . . point evidently to physical peculiarities in certain parts of the Sun's body more favourable than in others to the production of the spots, on the one hand; and on the other, to a general influence of its rotation on its axis as a determining cause in their distribution and arrangement, and would appear indicative of a system of movements in the fluids which constitute its luminous surface; bearing no remote analogy to our trade-winds—from whatever cause arising." More often than not when a main spot has a train of minor spots as followers that train will be found extending eastwards from the east side of the spot, rather than in any other direction.

Spots remain visible for very diverse lengths of time, from the extreme of a few minutes up to a few months; but a few days up to, say, one month, may, in a general way, be suggested as their ordinary limits of endurance. As the Sun rotates on its axis in 25¼ days, and as the spots may be said to be, practically speaking, fixed or nearly so with respect to the Sun's body, no spot can remain continuously visible for more than about 12½ days, being half the duration of the Sun's axial rotation.

With regard to their size, spots vary as much as they do in their duration. The majority of them are telescopic, that is, are only visible with the aid of a telescope; but instances are not uncommon of spots sufficiently large to be visible to the naked eye. The ancients knew nothing about the physical constitution of the Sun, and their few allusions to the subject were mere guesses of the wildest character. They were, however, able to notice now and then that when the Sun was near the horizon certain black spots could sometimes be distinguished with the naked eye, but they took these for planets in conjunction with the Sun, or phenomena of unknown origin. Earliest in point of date of those who have left on record accounts of naked eye sun-spots are undoubtedly the Chinese. In a species of Cyclopædia ascribed to a certain Ma-touan-lin (whose records of comets have been of the greatest possible use to astronomers), we find an account of 45 sun-spots seen during a period of 904 years, from 301 A. D. to 1205 A. D. In order to convey an idea of the relative size of the spots, the observers compared them to eggs, dates, plums, etc., as the case might be. The observations often extended over several days; some indeed to as many as ten consecutive days, and there seem no grounds for doubting the authenticity of the observations thus handed down to us. A few stray observations of sun-spots were recorded in Europe before the invention of the telescope. Adelmus, a Benedictine monk, makes mention of a black spot on the Sun on March 17, 807. It is also stated that such a spot was seen by Averroes in 1161. Kepler himself seems to have unconsciously once seen a spot on the Sun with the naked eye, though he supposed he was looking at a transit of the planet Mercury. None of these early observers have told us the way in which they made their observations, but the smallest of boys who has any claim to scientific knowledge is aware of the fact, that by the use of so simple an expedient as a piece of glass blackened with smoke, spots which are of sufficient size can be seen with the naked eye. Before telescopes came into use it was customary to receive the solar rays in a dark chamber through a little circular hole cut in a shutter. It was thus that J. Fabricius succeeded in December 1610 in seeing a considerable spot and following its movement sufficiently well to enable him to determine roughly the period of the Sun's rotation.

The spots may often be easily observed with telescopes of small dimensions, taking care, however, to place in front of the eye-piece a piece of strongly-coloured glass. For this purpose glasses of various colours are used, but none so good as dark green or dark neutral tint. It is not altogether easy to say positively how large a spot must be for it to be visible with the naked eye, or an opera glass, but probably it may be taken generally that no spot of lesser diameter than 1′ of arc can be so seen. This measurement must be deemed to apply to that central portion of a normal spot already mentioned as being what is called the nucleus, because penumbræ may be more than 1′ in diameter without being visible to the naked eye, for the reason that their shading is so much less pronounced than the shading of umbræ. Very large and conspicuous spots are comparatively rare, though during the years 1893 and 1894 there were an unusual number of such spots. It often happens that a conspicuous group is the result of the merging or joining up of several smaller groups. In such cases a group may extend over an area on the Sun 3′ or 4′ of arc in length by 2′ or 3′ in breadth. The largest spot on record seems to have been one seen on September 30, 1858, the length of which in one direction amounted to more than 140,000 miles.

The observation of spots on the Sun by projecting them on to a white paper screen with the aid of a telescope is a method so convenient and so exact as to deserve a detailed description, the more so as it is so little used. Let there be made in the shutter of a darkened room a hole so much larger than the diameter of the telescope to be used as will allow a certain amount of play to the telescope tube, backwards and forwards, up and down, and from right to left. Direct the telescope to the Sun and draw out the eye-piece to such a distance from the object-glass as that the image projected on a white screen held behind may be sharply defined at its edges. If there are any spots on the Sun at the time they will then be seen clearly exhibited on the screen. An image obtained in this way is reversed as compared with the image seen by looking at the Sun through a telescope directly. If therefore the telescope is armed with the ordinary astronomical eye-piece, which inverts, then the projection will be direct, that is to say, on the screen the N. S. E. and W. points will correspond with the same terrestrial points. Under such circumstances the spots will be seen to enter the Sun's disc on the E. side and to go off on the W. side. The contrary condition of things would arise if a Galilean telescope or a terrestrial telescope of any kind were made use of. These instruments erect the image, and therefore will give by projection a reversed image, in which we shall see the spots moving apparently in a direction contrary to their true direction.

If the reader has grasped the broad general outlines now given respecting the Sun and its spots he will perhaps be interested to learn a few further details, but these must be presented in a somewhat disjointed fashion, because the multitude of facts on record concerning sun-spots are so great as to render a methodical treatment of them extremely difficult within the limits here imposed on me. These matters have been gone into in a very exhaustive way by Secchi in his great treatise on the Sun, and in what follows I have made much use of his observations.

Let us look a little further into the laws regulating the movement of the spots. If it is not a question of seeing a spot spring into view, but of watching one already in existence, we shall, in general, see such a spot appear on the Eastern limb of the Sun just after having turned the corner, so to speak. The spots traverse the Sun's disc in lines which are apparently oblique with reference to the diurnal movement and the plane of the ecliptic, and after about 13 days they will disappear at the Western limb if they have not done so before by reason of physical changes in their condition. It is not uncommon for a spot after remaining invisible for 13 days on the other side of the Sun, so to speak, to reappear on the Eastern limb and make a second passage across the Sun; sometimes a third, and indeed sometimes even a fourth, passage may be observed, but more generally they change their form and vanish altogether either before passing off the visible disc, or whilst they are on the opposite side as viewed from the Earth.

When several spots appear simultaneously, they describe in the same period of time similar paths which are sensibly parallel to one another although they may be in very different latitudes. The conclusion from this is inevitable, that spots are not bodies independent of the Sun, as satellites would be, but that they are connected with the Sun's surface, and are affected by its movement of rotation. If we make every day for a few days in succession a drawing of the Sun's disc with any spots that are visible duly marked thereon, we shall see that their apparent progress

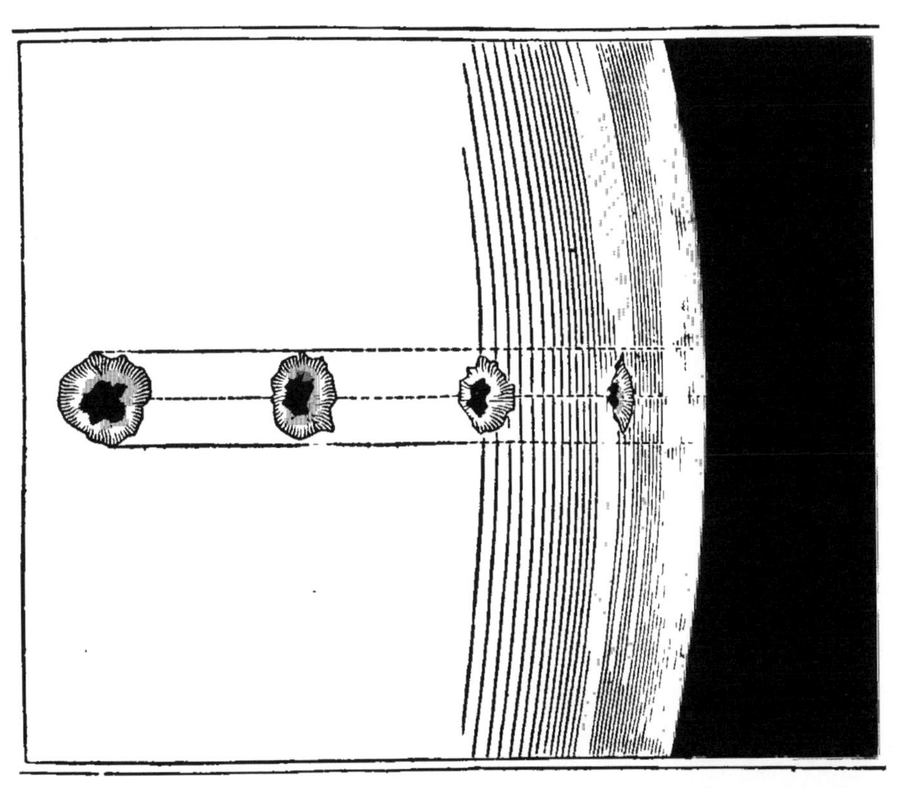

FIG. 6.—Change of Form in Sun-spots owing to the Sun's rotation.

is rapid near the centre of the Sun, but slow near either limb. These differences, however, are apparent and not real, for their movement appears to us to take place along a plane surface, whilst in reality it takes place along a circle parallel to the solar equator. The spots in approaching the Sun's W. limb, if they happen to seem somewhat circular in form when near the centre, first become oval, and then seem to contract almost into mere lines. These changes are simple effects of perspective, and are to be explained in the same manner as the apparent decrease in the size of many of the spots is often explicable. But this condition of things proves, however, that the spots belong to the actual surface of the Sun, for, on a contrary supposition, we should have to regard them as circular bodies greatly flattened like lozenges, and this would be contrary to all we know of the forms affected by the heavenly bodies. Of course besides the apparent changes of form just alluded to as the effect of perspective, it is abundantly certain that solar spots often undergo very real changes of form, not only from day to day, but in the course of a few hours. Several spots will often become amalgamated into one, and it was ephemeral changes of this character which hindered generally the early observers from determining with precision the duration of the Sun's rotation.

The apparent movements of the spots vary also from month to month during the year according to the season. In March their paths are very elongated ellipses with the convexity towards the N., the longer axis of the ellipse being almost parallel to the ecliptic. After that epoch the curvature of the ecllipse diminishes gradually, at the same time that the major axis becomes inclined to the ecliptic, so that by June the flattening of the ellipse has proceeded so far that the path has become a straight line. Between June and September the elliptical form reappears but in a reversed position; then, following these reversed phases, the ellipticity decreases, and for the second time there is an epoch of straight lines. This happens in December, but the straight lines are inclined in a converse direction to that which was the case in June. It must again be impressed on the reader that all these seemingly different forms of path pursued by the spots are merely effects of perspective, for in reality, the spots in crossing the Sun's disc describe lines which are virtually parallel to the solar equator. These projections really depend of course on the position of the observer on the Earth, and vary as his position varies during the Earth's annual circuit round the Sun. The number of the spots varies through wide limits. Sometimes they are so numerous that a single observation will enable us to recognise the position of the zones of maximum frequency. Sometimes, on the other hand, they are so scarce, that many weeks may pass away without hardly one being seen. A remarkable regularity is now recognised in the succession of these periods of abundance and scarcity, as we shall see later on.

It is both useful and interesting in studying the spots to record methodically their number and their size, but it is not easy to teach observers how to do this so systematically that observations by one person can be brought into comparison with those of another. Photography and hand-drawing on a screen alone furnish a trustworthy basis of operations. Spots in general may naturally be classified into (1) isolated spots or points, and (2) groups of spots; but often one observer will describe as a small spot an object which another observer would regard as a mere point; and one observer will record several groups where another observer will see but one. A very few days' experience with a telescope will bring home to the observer's mind the difficulty of dealing with the spots where it is a question of systematic methodical observation of them.

Let us now take a brief survey of some of the theories which have been put forth regarding the nature of the spots on the Sun. In the early days of the telescope, that is to say, during the 17th century, two general ideas were current. Some thought the spots to be shapeless satellites revolving round the Sun; others that they were clouds, or aggregations of smoke, floating about in a solar atmosphere. Scheiner, the author of the first theory, abandoned it towards the close of his life, having arrived at the conclusion that the spots were situated below the general level of the Sun's surface. Another idea, but of later date, was that the Sun is a liquid and incandescent mass of matter, and the spots immense fragments of *Scoriæ*, or clinkers, floating upon an ocean of fire.

Somewhat more than a century after the spots had been generally studied with the aid of a telescope a Scotchman named Wilson made a memorable discovery. He showed by the clearest evidence that they are cavities, and he propounded the first intelligible idea of the true physical constitution of the Sun, when he compared to a strongly illuminated cloud the luminous layer of solar material which we now term the "photosphere." On November 22, 1769, he observed on the Sun's disc a fine round spot encompassed by a penumbra, also circular, and concentric with the nucleus. He watched that spot up to the time that it disappeared, and he soon remarked that the penumbra ceased to be symmetrical: the part turned towards the centre of the Sun became smaller and smaller, and eventually disappeared altogether; whilst the part on the opposite side preserved its fulness and dimensions almost unchanged. Let us suppose we chanced to turn a telescope on to the Sun on a given day, and were fortunate enough to discover a spot in the centre of the disc, with a penumbra concentric with the nucleus. When such a spot arrives about midway towards the limb, it will exhibit a penumbra narrower on the left side than on the right; later on the penumbra will disappear almost or quite completely on the left side: then the nucleus itself will seem to be encroached upon. Finally, very near the limb, there will remain only a slender thread of penumbra, and the nucleus will have ceased to be directly visible. Such were the phases of transformation observed by Wilson and often studied since. Wilson suspected that he had come upon some great law that was ripe for disclosure, and in order not to be misled he waited for the return of the same spot, which indeed reappeared on the Sun's W. limb after about 14 days. Then he found himself face to face with the same phases reproduced, but in the reverse order: the penumbra contracted on one side and full on the other, widening out on the contracted side as the spot came up to the Sun's centre. Henceforth doubt was no longer possible; the spot had sensibly preserved the same shape during its passage, and the alterations noticed were only apparent, and resulted from an effect of perspective which was easy to be understood. The different phases presented by such a spot as that just spoken of will be so much the more sensible according as the depth of the cavity is greater; but if the depth is inconsiderable the bottom of the cavity will only disappear when a very oblique angle is attained, and this cannot happen except when the spot is very near to the limb. By observations carefully made under such circumstances it will be possible to determine the depth of the cavity, and Wilson found that the depth of a spot often amounted to about one-third of the Earth's radius. Wilson's theory was not accepted without dispute; it was contested by several astronomers, and in particular by Lalande. It was however taken up by Sir W. Herschel, and as modified by him has met with general acceptance down to the present time; though now and again challenged, perhaps most recently and most vehemently by Howlett, a sun spot observer of great experience. Wilson's discovery was the point of departure for the grand labours of Sir W. Herschel in the field of Solar Physics. Man of genius that Herschel was, he was above all things an observer who took his own line in what he did. He saw so many phenomena with the powerful instruments constructed by himself, he described so minutely the marvels which were revealed to him, that he left comparatively little for his successors to do so far as regards mere telescopic observation. Herschel's main idea as to the Sun was based on Wilson's discovery. He remarked with reason, as that astronomer had done, that if the spots are cavities the luminous matter could neither be properly called liquid nor gaseous; for then it would precipitate itself with frightful rapidity to fill up the void, and that would render it impossible that the spots should endure as we often see they do during several revolutions of the Sun. Moreover, the proper movements of the spots prove that the photosphere is not solid. We can therefore only liken it to fogs or clouds, and it must be suspended in an atmosphere similar to ours. Such is, according to Herschel, the only hypothesis which can explain the rapid changes which we witness. We shall see a little later on that these phenomena do admit of another explanation.

In a second memoir Herschel followed up this inquiry with an acuteness worthy of his genius. Unfortunately he allowed himself to be carried away with the idea that the Sun was inhabited in order to sustain this theory. He needed a solid kernel upon which his imaginary inhabitants could dwell; and also a means whereby he could protect them from the radiations of the photosphere. With this idea in view he conjectured the existence above the Sun's solid body of a layer of clouds always contiguous to the photosphere which enveloped it, and which always being rent when the photosphere was rent, thus enabled us to see the solid body of the Sun lying behind. These notions can only be described as very arbitrary, as unsupported by observation, and as involving explanations quite out of harmony with the principles of modern physics. However, the labours of Herschel resulted in so many positive discoveries of visible facts, and in so many just conclusions, that they contributed greatly to the growth of our present knowledge of the true constitution of the Sun.

Since Wilson's time, as Secchi pointedly remarks, astronomers generally have verified his observations with good instruments, and by an investigation of a great number of spots. De La Rue, discussing the Kew observations, found that of 89 regular spots 72 gave results which conformed to Wilson's ideas, whilst the remaining 17 were opposed thereto. There is nothing surprising in the existence of a contrarient minority when we consider the great changes which in reality often occur in the forms of the spots. De La Rue suggested a very simple expedient for showing that the spots are cavities. Take two photographs of the Sun made at an interval of one day: during that time every point on the Sun's surface will have been displaced, so far as the telescope is concerned, by about 15°. Place these photographs in a stereoscope, and we shall readily see the interior cavity, the edges of which will appear raised above the photosphere. It is impossible therefore to entertain the least doubt as to the truth of the theory that the spots are excavations in the luminous stratum which envelopes the whole of the solar globe.

If it be true that a spot is a cavity, it follows that when it reaches the margin of the solar disc we ought to detect a hollow place; and this will be so much the more easy to observe according as the cavity is larger and deeper. As a matter of fact, numerous observations of this sort have been recorded from the time of Cassini down to the present time under the designation of "notches" on the Sun's limb. On July 8, 1873, Secchi observed such a notch 8″, or 3600 miles deep.

Faye and some other astronomers are disposed to support a theory according to which the spots are nothing else than aerial cyclones, but this does not seem admissible. If the fundamental principle of a spot is that it arises from a whirling movement, the rays (so to speak) which compose the penumbræ must always be crooked, or the theory falls to the ground. It is quite true that indications of cyclonic action do sometimes appear, but they are at any rate very rare, for only a small percentage exhibit in a distinct manner a spiral structure. Moreover, when such a structure is seen it does not endure for the whole lifetime of the spot but only for a day or two: the spot may last a long time after it has lost its spiral features, if it ever had any. Sometimes even the

FIG. 7.—Sun-spot seen as a Notch.

whirling movement, after having slackened, begins again, but in the contrary direction. Under these circumstances, though this occasional spiral structure is very curious and interesting, we are not justified in taking it as the basis of a theory which has any pretensions to explain the general nature of sun-spots.

When we examine the Sun with instruments of large aperture and high magnifying power, we notice that its surface is far from being as smooth and uniform as it appears in a small telescope. On the contrary, it presents an irregular undulating appearance like a pond or other sheet of water agitated by the wind. Careful scrutiny with a powerful eye-piece reveals the fact that the Sun's surface is marked by a multitude of wrinkles and irregularities which it is well-nigh impossible to describe in words. More or less everywhere there is a general mottling visible; it is more distinct in some places than others, and especially so towards the centre of the disc. This peculiar appearance varies very much from time to time, and its distinctness seems to depend a great deal on the state of the Earth's atmosphere, for it becomes invisible when the air is disturbed; but these variations depend also on real variations of the photosphere—a fact which observations made in very calm weather are thought clearly to indicate.

It is often said that the Sun exhibits a granulated structure. If we wish to realise in the most precise manner what is meant by the word "granulation" as applied to the structure of the Sun, we must abandon the method of projection and examine the Sun directly with a powerful eye-piece, taking advantage of a moment when the atmosphere is perfectly calm, and before the eye-piece has had time to get hot. It may then be seen that the Sun's surface is covered with a multitude of little grains, nearly all of about the same size, but of different shape, though for the most part more or less oval. The small interstices which separate these grains form a net-work which is dark without being positively black. Secchi considered it difficult to name any known object which exactly answers in appearance to this structure, but he thought that we can find something resembling it in examining with a microscope milk which has been a little dried up, and the globules of which have lost their regular form. Exceptionally good atmospheric conditions are under all circumstances indispensable for the study of these details.

In point of fact, there is a mysterious uncertainty about the normal condition of the Sun's surface, in a visual sense, which a few years ago engendered a very vehement controversy, and led to the use of such expressions as "willow leaves," "rice grains," "sea beach," and "straw thatching," to indicate what was seen. All these words are too precise to be quite suitable to be taken literally, but perhaps, on the whole, "rice grains" is not altogether a bad expression to recall what certainly seems to be the granular surface of the Sun as we see it.

By making use of moderate magnifying powers, what we see will often convey the impression of a multitude of white points on a black net-work. This is very apparent during the first few moments that the telescope is brought to bear on the Sun, but its clearness quickly passes away because the eye gets fatigued, and the lenses becoming warm the air in the telescope tube gets disturbed because also warmed. Sometimes the appearance is a little different from that just described, and along with the white and brilliant points little black holes are intermixed. Oftentimes the grains appear as if suspended in a black net-work and heaped together in knots more or less shaded and more or less broad. Sometimes the grains exhibit a very elongated form, especially in the neighbourhood of the spots. It is these elongated forms to which Nasmyth applied the term "willow leaf," whilst Huggins thought "rice grains" a very suitable expression.

This granular or leaf-like structure—call it what we will—cannot be made out except with considerable optical assistance, for the grains being intrinsically very small, diffraction in enlarging them and causing them to encroach on one another necessarily produces a general confusion of image. The real dimensions of these grains cannot therefore readily be determined by direct measurement, but by comparing them with the wires used in micrometer eye-pieces it has been thought that their diameters may usually be regarded as equal to $\frac{1}{4}$ or $\frac{1}{8}$ of a second—say from 120 to 150 miles. The granules seem to be possessed of sensible movement, but presumably it is not always or even generally a movement of translation from place to place; only an undulatory movement like that of still water when a stone is cast into it. Nevertheless, probably in certain cases the granules actually are affected by a motion of translation, for in the vicinity of spots they may sometimes be seen flowing over the edges of the penumbræ. In order to explain the existence of the granules the strangest theories have been broached. Sir William Herschel having observed the granulations, applied to them the term "corrugations" or "furrows"—words somewhat inexact, perhaps, but by which, as his descriptions clearly show, he meant to designate the features which I am now treating of. He even noticed the dark network which separates the grains, and he applied to it the word "indentations."

These granulations are without doubt prominences, probably of hydrogen gas, which rise above the general surface, for this structure is much more sharp and distinct at the centre of the sun's disc than at the limbs; that is to say, near the limbs of the Sun they partially overlap one another, as indeed Herschel remarked. The idea of flames would satisfy these appearances: and as the spectroscope suggests to us that the Sun is habitually covered over with a multitude of little jets of flame, the observations which have been made compel the opinion that the grains are the summits of those prominences which exist all over the Sun's surface.

The surface is sometimes so thickly covered over with these granulations—the network is so conspicuous—that we can readily imagine that we see everywhere pores and the beginnings of spots, but this aspect is not permanent, and seems to depend to some extent on atmospheric causes combined also with actual changes in the Sun's surface itself. There seems however no doubt that the joints, so to speak, of the dark network already referred to do sometimes burst asunder and develope into spots.

The circumstances which accompany the formation of a spot cannot readily be specified with certainty. It is impossible to say that there exists any law as to this matter. Whilst some spots develope very slowly by the expansion of certain pores, others spring into existence quite suddenly. Yet it cannot be said that the formation of a spot is ever completely instantaneous however rapid it may be. The phenomenon is often announced some days in advance: we may perceive in the photosphere a great agitation which often manifests itself by some very brilliant *faculæ* (to be described presently) giving birth to one or more pores. Very often we next notice some groups of little black spots, as if the luminous stratum was becoming thinner in such a way as to disappear little by little and leave a large black nucleus uncovered. At the commencement of the business there is usually no clearly defined penumbra. This developes itself gradually and acquires a regular outline, just as the spot itself often takes a somewhat circular form. This tranquil and peaceable formation of a spot only happens at a time when calm seems to reign in the solar atmosphere: in general the development is more tumultuous and the stages more complicated.

As a rule a spot passes through three stages of existence:—(1) the Period of birth; (2) a Period of calm; and (3) the Period of dissolution. When a spot is on the point of closing up, the flow of the luminous matter which it, as it were, attracts, is not directed uniformly towards the centre; it seems that the photospheric masses, no longer meeting with resistance, are precipitated promiscuously anywhere so as to fill up the hole. It is impossible to describe in detail the phases which irregular spots go through, but two things may always be remarked: that their structure is characterized by the existence of luminous filaments, and that these filaments converge towards one or several centres.

Secchi thus sums up certain conclusions which he arrived at relating to spots generally:—(1) It is not on the surface of any solid body that the solar spots are manifested; they are produced in a fluid mass, the fluidity of which is represented by a gas, so that the constitution of this medium may be likened to that of flames or clouds; (2) the known details respecting the constitution of the penumbra and the phenomena exhibited prove that the penumbra is not a mass of obscure matter which floats across luminous matter, but that it is on the contrary a case of luminous matter invading and floating about over darker materials and so producing a half tint.

All the available evidence which we possess may be said to show that the spots are not merely superficial appearances, but that they have their origin deep in the interior of the Sun, and are produced by the operation of causes still unknown to us which affect and disturb the Sun's mass to an extent which is sometimes very considerable. The spots then are only the results of a great agitation in the materials of which the Sun is composed, and this agitation extends far down below the limits of the visible dark nucleus whatever that may consist of.

Besides the spots, streaks of light may frequently be remarked upon the surface of the Sun towards the margin of the disc. These are termed *faculæ* (torches), and they are often found near the spots, or where spots have previously existed or have afterwards appeared. When quite near the Sun's limb these faculæ are usually more or less parallel to the limb. They are of irregular form and may be likened to certain kinds of coral. They generally appear to be more luminous than the solar surface immediately adjacent to them, but it is not improbable that this is an optical illusion depending upon the fact that the edges of the Sun always appear much more luminous than the centre. This last-named fact may be readily recognised by the employment of a high magnifying power, and moving the telescope rapidly from the limb to the centre of the disc. If the Sun be projected on a screen, as already mentioned, this degradation of the Sun's light from centre to circumference becomes particularly manifest.

After having studied the structure and the movement of the spots, one is naturally led to ask if their apparitions at different periods are subject to any general law. This question is one which has much engaged the attention of modern astronomers. The older observers noticed that the number of the spots visible differed in different years. There were said to have been periods when months and even years passed away without any spots being observed. Even allowing that this statement, so far as "years" are concerned, might be exaggerated, and that the absence of spots was due to the want of sufficient care in making the observations, and especially to the want of efficient instruments, it is none the less true that the number of the spots is extremely variable, and that there have been epochs when they were very scarce.

Sir W. Herschel was the first who devoted himself to the question of seeking to establish a relation between the variation of the spots and terrestrial meteorology. For the want of any better object, he compared the annual number of the spots with the price of wheat; but it is easy to see that nothing could result from such a comparison. Without doubt the meteorological phenomena of the globe must depend to some extent on solar changes: but the term of comparison selected by Herschel had no direct bearing on the state of the Sun.

In our time this question has been investigated to its very foundation by Wolf, Director for many years at the Observatory of Zurich. It is to his zeal that we owe a very interesting assemblage of old observations which were buried in archives and chronicles. It was he who endeavoured to reduce them into a systematic form, so as to supply as far as possible the numerous gaps which exist in the different series.

The two most attentive observers at the period when the spots were discovered were Marriott at Oxford and Scheiner at Ingoldstadt, but Scheiner himself has informed us that he did not note down all the spots which he saw; he only recorded those which were likely to assist him in his special task of determining the period of the Sun's rotation. Several observers after him made isolated series of observations; but some of these have been lost and the others show important gaps. J. G. Staudacher, at Nuremburg, observed the Sun with great perseverance during fifty years from 1749 to 1799. Before him the Cassinis, Maraldi, and others were engaged in the same sort of work, but only in an indirect way: that is to say, they contented themselves, whilst making meridional observations of the Sun, with noting anything in the way of spots which they deemed important. Zucconi and Flaugergues also left behind them a good collection of observations which Wolf utilised, rendering them comparable one with another by applying suitable corrections. The great difficulty herein arises from the fact that the observers were not provided with instruments of equal power; one man, armed with a better telescope than his contemporaries, naturally observed and recorded spots which would escape the others. The numbers entered in their registers are therefore not comparable *inter se.* Wolf endeavoured to replace these numbers by others which would represent the spots which might have been seen if the observers had all employed telescopes of a given kind and power. The result of his efforts in this direction is an almost continuous series of Sun-spot records from an epoch sufficiently remote, up to the time when this branch of science was taken up with the vigour of modern scientific methods.

The observer who most assiduously devoted himself to this subject in modern times was Schwabe of Dessau. From 1826 to 1868 he never failed to make daily observations when the weather permitted him. His series of records is specially valuable, for Carrington's fits in with it, and with that in turn Sporer's is comparable, and the chain is complete by the later photographic and other observations. All these Sun-spot records, though differing in their details, may easily be used together when it is a question of working out relative annual fluctuations.

At the present time there are many Astronomers who are engaged in observing the spots with care; but just as formerly there are few who possess sufficient perseverance. The photographic method is excellent, but it takes much time and is costly. Some have decried, in a very unreasonable manner, a drawing made by hand: such a drawing, of sufficient size, and executed by projection by a skilful draughtsman with a telescope driven by clockwork, may stand comparison with a photograph, and this method has a better chance of being persevered in. The Rev. F. Howlett's name must be mentioned in this connection as a draughtsman who has accomplished much by hand drawing. Though the once famous Kew observations have been discontinued, they have been replaced by a new series at Greenwich with similar appliances; whilst Janssen at Meudon has also been carrying on for a number of years a splendid course of photographic records.

Schwabe, when he had collected a considerable number of observations, recognised clear indications of periodicity. Very definite epochs of maxima and minima succeeded one another at intervals of 10 or 11 years. It is true that in following out such a study the observations are certain to be in a sense a little defective. At first it was not possible to observe the Sun every day, and the gaps which resulted from bad weather necessarily added to the number of days which had to be set down as being without spots. Moreover, every method of numbering the spots must be a little arbitrary: there are often groups which, in consequence of their sub-divisions, may be counted in different ways: but in a mass of observations so considerable as those of Schwabe's, such uncertainties will compensate for one another and will disappear in the final result. In fact the law is so striking that it suffices to cast one's eye over his table* to see that.

That table is both interesting and instructive at the same time. The numbers exhibited in it speak for themselves, and it is sufficient to exam-

---

* Given in full in my *Handbook of Astronomy*, 4th ed., vol. i., p. 26.

4

ine them with even a small amount of attention to realise the certainty of the conclusions which have been drawn.

It is therefore now to be deemed an ascertained fact that there are periodical maxima and minima in the display of spots, and that the extent of the period is between 10 and 12 years. In order to determine this value with the utmost exactness, some astronomers have had recourse to early observations. Wolf of Zurich made this the subject of some very interesting inquiries. He was able to establish the chronology of the phases which the Sun has passed through from the time of the first discovery of the spots to the present day—more than $2\frac{1}{2}$ centuries. His calculations led him to a period of $11\frac{1}{9}$ years. Lamont fixed upon 10.43 years, but this number does not represent the more recent observations with sufficient precision.

In order to exhibit this law in the plainest possible manner the dates of maxima and minima should be laid down on ruled paper in proper mathematical form, the *abscissæ* of the curve representing the years, and the *ordinates* the number of spots observed.

An examination of a curve thus plotted shows two things:—(1) That the period is clearly an eleven-year one, as has been already stated; (2) that it is not however quite as simple in its form as it was at first thought to be; for in reality there are two periods superposed, the one rather more than half a century long, and the other extending over the 11 years already spoken of. We do not possess early observations sufficiently numerous and sufficiently good to enable us to draw any unimpeachable conclusions as to the nature of the long period; we can only be certain that it exists. The later labours of Wolf, however, fixed that period at 55½ years. It is a result of this that, according to Loomis, a period of comparative calm on the Sun existed between 1810 and 1825.

Each maximum lies nearer to the minimum which precedes it than to the minimum which follows it, for the spots increase during 3.7 years, and then diminish during 7.4 years. According to De La Rue the increase occupies 3.52 years, and diminution 7.55 years. This concurrence between De La Rue and Wolf is surprising considering the diversity of the methods which led to results almost identical, the one set being based on the number of the spots, and the other on the superficial extent of the spots. The different periods in succession are not absolutely identical: but it has been remarked that if during any one period the decrease is retarded or accelerated, then the increase next following will be lengthened or contracted to a corresponding extent. In consequence of this we are sometimes able to predict with fair accuracy when the next ensuing maximum or minimum will take place.

The most striking feature of such a curve as that just alluded to is the very sensible secondary augmentation which happens very soon after the principal maximum.

A very curious circumstance has come to light in connection with the epochs of maxima and minima. In arranging the spots according to their latitude and longitude on a diagram sufficiently contracted, Carrington found that their latitude decreases gradually as the period of minimum draws near; then when their number begins to increase they begin to appear again at a higher latitude. This seems to be a definite law. At any rate Carrington's conclusion has been found to hold good by the observations of Sporer and Secchi.

The variations of the spots which we now recognise naturally recall those obscurations of the Sun which are recorded in history; but it is necessary to accept many of these with caution. A great number of these phenomena which attracted the attention of people in early times are only eclipses badly observed and still more badly described. In other instances the obscuration has been produced by very protracted dry fogs. It is probably to this last-named cause that we must ascribe the obscuration which, according to Kepler and Gemma Frisius, took place in 1547.

It was in some such way as this that, according to Virgil (Georg. i, 630), who has echoed a tradition which he found in history, the Sun was obscured at the death of Cæsar :—

> Ille etiam extincto miseratus Cæsare Romam
> Quum caput obscura nitidum ferrugine texit,
> Impiaque æternam timuerunt sæcula noctem.

In the year 553 A. D., and again in the year 626 A. D. the Sun remained obscured for several months; but these facts (if facts they are) besides being ill-observed, and clothed, no doubt, in extremely exaggerated language, are brought to our notice as having occurred at epochs which are quite independent of one another, whilst the variations in the markings on the Sun, which we have just been talking about, present an almost mathematical regularity of sequence.

We must now institute some inquiries as to

the causes of the periodicity of the spots. A periodicity so well established would naturally invite astronomers to seek the causes which produced it. The presence of spots only in the Zodiacal regions led Galileo to suspect the existence of some relation between the spots and the position of the planets; but there is in this a mere surmise, which, when it was made, had nothing to justify it, and it is still impossible for us to say anything for certain on the point. The determining cause of the periodicity may exist in the interior of the Sun, and may depend on circumstances which will for ever remain unknown to us. Or it may be something external: it may be due after all to the influence of the planets. It remains for us, therefore, to search and see if any such influence can be traced.

According to Wolf, the attraction of the planets, or of some of them, is the real cause of the periodicity which we are dealing with; that attraction producing on the surface of the solar globe true tides, which give birth to the spots, these tides themselves experiencing periodic variations owing to the periodic changes of position of the celestial bodies which cause them. It has even been thought safe to assert that the fact of the principal period coinciding with the revolution of Jupiter is of momentous significance; but this coincidence seems purely accidental, and no certain conclusion can be drawn as to this matter. The influence of Mercury and Venus would perhaps be much more potent, for their distance from the Sun is not very great, and this should render their influence more sensible. On the other hand, their masses appear to be too small to be capable of producing any sufficient effect.

De La Rue, Balfour Stewart, and Lowy most perseveringly studied this point of solar physics. They seem to have arrived at the conclusion that the conjunctions of Venus and Jupiter do exercise a certain amount of influence on the number of the spots and on their latitude; and that this influence is less considerable when Venus is situated in the plane of the solar equator. At any rate it is a fact, that a great number of the visible inequalities in a duly plotted curve of the spots do really correspond to special positions of these two planets.

In order to determine with more precision these coincidences and the importance which attaches to them, De La Rue extended his inquiries. He separately analysed many different groups of spots, selecting for his purpose more particularly those of which the observations happened to have been specially continuous and complete, giving a preference moreover to those which had been observed in the central portions of the Sun's disc. From an investigation of 794 groups De La Rue arrived at the following conclusions:— (1) If we take a meridian passing through the middle of the disc and represented by a diameter perpendicular to the equator, we find that the mean size of the spots is not the same with regard to that meridian. It appears certain that the correction required for perspective does not suffice to explain this difference; and that another element must be introduced in order to secure that the apparent dimensions of the spots may be the same on both sides. We do not yet possess a very clear explanation of this fact; but the most probable is this:—the spots are surrounded by a projecting bank, which seems to disappear in part during their transit across the Sun. This bank is more elevated on the preceding than on the following side; accordingly, the spots ought to seem smaller when they are in the eastern half of the disc; larger when they are in the western half; for in the first position the observer's eye meets an elevated obstacle, which hides a portion of the spot itself. (2) De La Rue specially studied the spots observed at the times when the planets Venus and Mars were at a heliocentric distance from the Earth equal to 0, 90, 180, and 270 degrees, and arrived at this result; the spots are larger in the part of the Sun which is away from Venus and Mars, and they are smaller on the side on which these planets happen to be. The same result was obtained, whether Carrington's figures or the Kew photographs were employed. (3) Meanwhile it does not appear that Jupiter emits any similar influence. This influence should be easily perceived, for if we calculate the action of the planets in the way that we calculate the tides, treating it as directly proportional to the masses and inversely proportional to the cubes of the distances, the influence of Jupiter should greatly outweigh that of Venus.

Wolf thought that he had noticed traces of some influence being exerted by Saturn; but this remains altogether without confirmation.

De La Rue noticed that large spots are generally situated at extremities of the same diameter. This law also often applies to the development of large prominences. The coincidence agrees well with the theory that there exists on the Sun some action resembling that of our tides.

Whatever may be the amount of probability

which attaches to these explanations we ought not to forget that we are still far off from possessing the power of giving a vigorous demonstration of them. If we consider with attention the periodical variations of the spots we shall not be long in coming to the conclusion that it is impossible to connect them directly with any one astronomical function in particular, for the spots appear in a sudden and irregular manner which contrasts in a striking degree with the continuous and progressive action of the ordinary perturbations which we meet with in the study of Celestial Mechanics. There is but one reply possible to this objection. The spots and their changes must be visible manifestations of the periodical activity of the Sun—an activity which itself depends (as assumed) on the action of the planets and on their relative positions. The cause, thus defined, of the Sun's activity may be very regular; the activity itself may vary in a continuous manner without the resulting phenomena possessing the same continuity and the same regularity. We see this in the periodical succession of the Seasons on the Earth. The position of the Sun, and consequently its manner of acting upon our globe, varies with a remarkable uniformity, but nevertheless the meteorological phenomena which result are irregular and capricious. Thus it comes about that physicists are more and more inclined to believe that the spots are only secondary effects produced by causes more important and more fundamental.

Whatever may be our ignorance as to the causes which produce variations in the Sun's activity we may at least draw one conclusion from the preceding remarks: it is, that the Sun is a very long way from having arrived at a state of tranquillity and freedom from internal commotion. On the contrary, it is the seat of great movements. Its activity is subject to numberless periodical changes which ought in their turn to influence the intensity of the heat and light given out by the Sun; and so re-act on the planets which receive their heat, light, and life from the Sun.

No account of the periodicity of the spots on the Sun can be deemed complete which does not include information respecting certain other periodical phenomena which have been found to exhibit features of alternation closely resembling in their sequence and character the periodical changes which take place in regard to the spots on the Sun. There is evidently a deep mystery lying hid under the curious fact (which is clearly established) that the 11-year period of the spots coincides in a manner as unexpected as it is certain with the period of the variation of terrestrial magnetism. The magnetic needle is subject to a diurnal variation which reaches its extreme amount every 11 years, and not only so, but the epoch of maximum variation corresponds with the epoch of the maximum prevalence of Sun spots. And similarly years in which the needle is least disturbed are also years in which the Sun spots are fewest. Two other very curious discoveries have also been made which are in evident close connection with the foregoing. The manifestation of the Aurora Borealis and of those strange currents of electricity known as magnetic earth currents (which travel below the Earth's surface and frequently interfere with telegraphic operations), likewise exhibit periodical changes which take 11 years to go through all their stages. This fact alone would be sufficiently curious, but when we come to find that the curve which exhibits the changes these two manifestations of force go through, also shows that their maxima and minima are contemporaneous with the maxima and

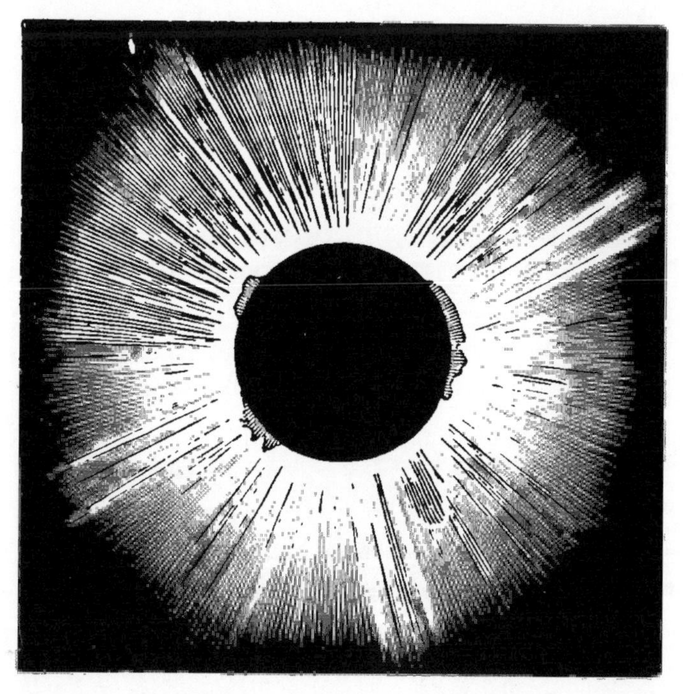

FIG. 8.—The Sun totally eclipsed, July 18, 1860 (Feilitzsch).

minima of the Sun spots and magnetic needle variations, we cannot doubt that (to use Balfour Stewart's words) "a bond of union exists between these four phenomena. The question next arises, what is the nature of this bond? Now, with respect to that which connects Sun spots with magnetic disturbances we can as yet form no conjecture." To cut a long story short, it may be said generally that whilst without doubt electricity is the common basis of the three last-named of the four phenomena just mentioned, it seems scarcely too great a stretch of the imagination to go one step further and suggest that electricity has in some or other occult manner something to do with all these things and therefore with the spots on the Sun.

The reader who has followed me thus far will by this time be in a position to appreciate a remark made in an earlier part of this chapter, that the multitude of facts known to us in connection with the Sun and its spots is so great, as to render it impossible to exhibit in a single chapter anything more than the barest outline of them. The numerous observations of recent eclipses of the Sun, especially since that of 1860, and the extensive application of the spectroscope to the Sun both in connection with these eclipses, and generally, may be said to have completely revolutionised our knowledge of solar phenomena during the present generation; or perhaps it might be more correct to say have enormously increased our knowledge of the facts of the case and have revolutionised in no small degree the conclusions deduced from the facts.

CHAPTER III.

MERCURY.

So far as we know at present, Mercury is the nearest planet to the Sun. The circumstances under which it presents itself to us and a brief general account of its movements have already been stated. In the present chapter, therefore (and this remark applies in substance to each of the succeeding chapters appropriated to particular planets), I shall limit myself to such topics as seem to be of interest to an observer armed with a telescope. Mercury, as already mentioned, exhibits from time to time phases which may be said to be the same as those of the moon; but as the only chance of seeing it is when it is at its greatest distance east or west of the Sun, practically it can only be studied when in, or rather near to, what may be called the half-moon phase; and even then observations on its physical appearance can only be obtained with difficulty. Perhaps its most definite feature is its colour. This, undoubtedly, is more or less pink. Strange to say, in spite of the multiplication of telescopes and observers, comparatively little attention has been paid to this planet, and we really know very little more about it than Schroter told us nearly a hundred years ago. He obtained what he conceived to be satisfactory evidence of the existence of at any rate one mountain, having a height of about 11 English miles—a height which it will be noted, far exceeds, not only relatively but absolutely, any mountain on the earth. What Schroter based this conclusion upon was the fact that when the planet was near inferior conjunction, the southern horn presented a truncated appearance, which might be the result of a lofty projection arresting the Sun's light. Schroter also announced that Mercury rotated on its axis in 24 hours 5 minutes. Sir W. Herschel failed to satisfy himself that Schroter's conclusions were well-founded, but it must certainly be admitted that some support for them is furnished by certain observations made within the last few years. It is matter for regret, however, that most of these were made with instruments of sizes which, for the most part, cannot be said to have been equal to the task to which they were applied. The truncature of the southern horn first spoken of by Schroter, was thought by Denning, in 1882, to be obvious; and in the same year, by watching the displacement of certain bright and dusky spaces on the disc, the same observer concluded that a rotation period of about 25 hours was indicated.

In 1882 Schiaparelli at Milan commenced a prolonged study of Mercury. Believing that it was essential to observe through a good condition of atmosphere, and that this was impossible if the planet were only looked at in twilight, when it was necessarily at a low altitude, Schiaparelli made all his observations with the Sun and planet high up in the heavens. He considered, in effect, that the blaze of the Sun's light was a lesser evil than the tremors inseparable from observations of the planet, clear it might be in some degree of inconvenient Sun-light, but viewed through the vapours and atmospheric disturbances, which always spoil all observations near the horizon. Schiaparelli's observations yielded various results, most of them novel, and one of them very startling. He considers Mercury to be a much spotted globe and to be enveloped in a tolerably dense atmosphere. He thought he noticed brownish stripes and streaks (which might be regarded as permanent markings), more clearly visible on some occasions than on others; and that these systematically disappeared near the limb, owing to the increased depth there of the atmosphere through which they had to be looked at.

The foregoing observations may be regarded as not unreasonable; they may even be accepted without further question. But what are we to say to Schiaparelli's conclusions that these markings are so nearly permanent, taking one day with another, that Mercury's rotation cannot be measured in hours at all, but is a matter of days,—in point of fact, of 88 days; and that in reality Mercury occupies in its rotation on its axis the whole of the 88 days which constitute its sidereal year, or period of revolution round the Sun. The counterpart of this for us would be that, instead of the inhabitants of the earth having a day of 24 hours, they would have only one day and night every 365 days. Astronomers are not at present satisfied to accept this conclusion in regard to Mercury.

Some observers have thought that Mercury is more easy to observe than Venus, and that, speaking generally, its surface, if we could only get to see it constantly under favourable circumstances, might be considered to resemble in most respects that of Mars. Mercury revolves round the Sun at a mean distance of 36 millions of miles. Owing, however, to the fact that the eccentricity of its orbit (or its departure from the circular form) is greater than that of any of the other major planets, it may approach to within 28½ millions of miles or recede to more than 43 millions of miles. Its apparent diameter varies between 4½″ in superior conjunction to 13″ in inferior conjunction. The real diameter may be taken at about 3000 miles.

CHAPTER IV.

VENUS.

THE planet Venus has two things in common with Mercury. One is, that being an inferior planet, that is to say, a planet revolving round the Sun in an orbit within that of the Earth, it is never very far distant from the Sun, and therefore can never be seen on a distinctly dark sky. The second point alluded to arises out of the first; Venus exhibits from time to time a series of phases which are identical in character with those of Mercury, and therefore with those of the Moon. Venus differs, however, from Mercury in the very important point of size. Inasmuch as its diameter is considerably more than double the diameter of Mercury it has a surface more than six times as great, and therefore exhibits a far larger area of illumination than Mercury does. The result of this (coupled with another fact which will be stated presently) is that the planet may often be easily seen in broad daylight, and sometimes casts a sensible shadow at night. Under special circumstances, which recur every 8 years, this planet shines with very peculiar brilliancy. True, that only about ¼th of the whole disc is then illuminated, but that fraction transmits to us more light than phases of greater extent do, because these latter coincide with epochs when the planet is more remote from the Earth.

Spots and shadings have on various occasions been noticed on Venus, and though it is not easy to harmonise the various accounts, there seems no doubt of the reality of the facts, or that they must be ascribed to the existence of mountains. Schroter found very much the same state of things to exist on Venus that he found on Mercury, and putting together what he saw he arrived at the conclusion that Venus possesses mountains of considerable height, and that his observations must be taken to imply that the planet revolved on its axis in rather more than 23 hours. This conclusion as regards the planet's axial rotation was not first arrived at by Schroter, for the two Cassinis, one about 1666, and the other about 1740, both ascribed to Venus a rotation period of about 23 hours, an evaluation which was fully confirmed by Di Vico at Rome between 1839 and 1841, and by Flammarion in 1894.

What has been already said with respect to Mercury is true also of Venus, namely that it has been much neglected by modern observers; and accordingly an announcement made by Schiaparelli in 1890, that the rotation period of Venus is to be measured not by hours but by months, came upon the astronomical world as a startling revelation; but it is a revelation which has been keenly contested, and certainly awaits legal proof. Schiaparelli has not ventured to assert as he has done in the case of Mercury, that Venus's rotation period is identical with the period of 7½ months in which it revolves round the Sun; he only claims this as a strong probability arising out of what he says he is certain of, namely that its period of rotation cannot be less than six months and may be as much as nine months. His assumption is that previous observers in endeavouring to ascertain Venus's rotation period have used and relied upon evanescent shadings which probably were of atmospheric origin and scarcely recognisable from day to day, whereas he fixed his attention upon round denned white spots, which, whatever their origin, are so far permanent that their existence has been spoken of for two centuries. Miss Clarke thus puts the matter:—"His steady watch over them showed the invariability of their position with regard to the terminator; and this is as much as to say that the regions of day and night do not shift on the surface of the planet. In other words she keeps the same face always turned towards the Sun."

Various recent observations, some of them made with the express object of throwing light upon Schiaparelli's conclusions, are strangely contradictory. Perrotin at Nice in 1890 thought his observations confirmed Schiaparelli's; on the other hand Niesten at Brussels considered that numerous drawings of Venus made by himself and Stuyvaert between 1881 and 1890 harmonised well with DiVico's rotation period of 23h. 21m. 22s.; which Trouvelot in 1892 only wished to increase to about 24 hours.

There is a general consensus of opinion that great irregularities exist on the surface of Venus. These are made specially manifest to us in connection with the terminator or visible edge of the planet seen as an illuminated crescent. If the planet had a smooth surface this line would at all times be a perfect and continuous curve, instead of which it is frequently to be noticed as a jagged or broken line. Observations to this effect go back as far as 1643, when Fontana at Naples observed this to be the condition of the terminator. La Hire, Schröter, Madler, Di Vico and many others down to the present epoch have noted the same thing. The fact that the southern horn of

Venus is constantly to be seen blunted is so well established as to admit of no doubt, and this blunting is commonly ascribed to the existence of a lofty mountain, to which Schroter ascribed a height of 27 miles. Whatever we may think as to the precise accuracy of this figure, it seems impossible to doubt the main fact on which it depends; whilst a Belgian observer, Van Ertborn, in 1876 repeatedly saw a point of light in this locality which he regarded as due to Sun-light impinging on a detached peak, adjacent valleys remaining in shadow. This effect is common enough in the case of the Moon, and is familiar to all who are in the habit of studying the Moon.

FIG. 9.—Venus, Dec. 23, 1885.

The existence on Venus of an atmosphere of considerable density and extent is well established. Proof of this is to be found in the marked diminution of the planet's brilliancy towards the terminator; and in the faint curved line of light which occasionally may be seen when the planet is near inferior conjunction. When so situated, so much of the planet itself as can be seen illuminated shows as a narrow radiant crescent of light, ending off in two points called indifferently cusps or horns. It sometimes happens, however, that from the point of each cusp there runs round to the other cusp a faint continuation of the crescent, resulting in the general appearance of the planet being that of a nearly uniform ring of light. There is no known way in which the Sun can illuminate so much more than the half of Venus so as to permit of a perfect circle being visible except by supposing that an atmosphere exists on the planet and refracts (or transmits by bending, as it were, round the corner) a sufficient amount of Sun-light to give rise to the appearance in question. Further proof of the existence of an atmosphere on Venus is obtainable on those very rare occasions when the planet is seen passing across the disc of the Sun —a phenomenon known as a "Transit of Venus." It then nearly always happens that a hazy nebulous ring of feeble light may be detected encompassing the planet's disc indicative of course of the fact that the Sun's rays are there slightly obstructed in reaching the eye of an observer on

FIG. 10.—Venus near conjunction as a thin crescent, Sept. 21, 1887 (Flammarion).

the Earth. Some observers scrutinising Venus when in transit have thought that they were able to obtain, by means of the spectroscope, traces of aqueous vapour on the planet, but the evidence of this does not appear to be altogether clear or conclusive.

Everybody may be presumed to be acquainted with the spectacle popularly known as "The Old Moon in the New Moon's Arms" whereby when the Moon is only about two or three days old and exhibits but a narrow crescent of bright light, yet the whole outline of the disc is traceable on the sky. A phenomenon analogous to this may often be seen in the case of Venus when near its inferior conjunction. With the Moon the cause is due to the reflection of Earth-light (so to speak) to the Moon, but that explanation seems inadequate in respect of Venus, because it is conceived that the amount of Earth-light available is altogether insufficient for the purpose. Many other explanations have been put forward including phosphorescence on the surface of Venus, electrical displays in the nature of terrestrial auroræ, and what not, but it must be frankly confessed that astronomers are all at sea on the subject.

The existence of snow at the poles of Venus has been suspected by observers of tried skill and experience such as Phillips and Webb, though the idea was first broached by Gruithuisen in 1813. Flammarion's observations during 1892 and the two following years are distinctly confirmatory of this idea. He adds that as both polar caps are visible at the same time the planet's axis cannot be much inclined to the plane of its orbit.

Compared with all the other planets the absolute brightness of Venus stands very high. Of course it must be understood that by this phrase "absolute brightness" no more is meant than its reflective power. Venus is what it is by virtue of its power of reflecting Sun-light; presumably it has no inherent brightness of its own. What its reflective power is was probably never more effectively brought under the notice of a human eye than on September 26, 1878, when Nasmyth enjoyed an opportunity of seeing Venus and Mercury side by side for several hours in the same field of view. He speaks of Venus as resembling clean silver and Mercury as nothing better than lead or zinc. Seeing that owing to its greater proximity to the Sun the light incident on Mercury must be some 3½ times as strong as the light incident on Venus, it follows that the reflective power of Venus must be very great. As a matter of fact it has been calculated to be nearly equal to newly fallen snow; in other words to reflect fully 70 per cent. of the light which impinges on it.

Venus has no satellite; this fact seems certain. Yet half a dozen or more observers between 1645 and 1768 discovered such a satellite; observed it; followed it! This startling mystery, as it really was, attracted some years ago the attention of a very careful Belgian observer, Stroobant, who examined in a most painstaking manner all the recorded observations. His conclusions were that in almost all cases particular stars (which he identified) were mistaken for a satellite. Where the object seen was not capable of identification, possibly it was a minor planet; whilst in one instance it was probable that it was Uranus which had been seen and regarded as a satellite of Venus.

Venus is perhaps the planet which has most impressed the popular mind. For the earliest illustration of this statement we must go as far back as Homer who makes two references to it in the *Iliad.* These, in Pope's version, run as follows:—

> "As radiant Hesper shines with keener light,
> Far beaming o'er the silver host of night."
> —xxii. 399 [318].

> "The morning planet told th' approach of light;
> And fast behind, Aurora's warmer ray
> O'er the broad ocean pour'd the golden day."
> —xxiii. 281 [226].

The phases of Venus were first discovered by Galileo and were made known to the world, or rather to Kepler, in a mystic sentence which has often been quoted:—

"*Hæc immatura, a me jam frustra leguntur—oy*"

"These things not ripe; at present [read] in vain [by others] are read by me."

The former sentence transposed becomes—

Cynthiæ figuras æmulatur mater amorum.

The mother of loves [Venus] imitates the phases of Cynthia [the Moon].

Venus revolves round the Sun in 224½ days at a mean distance of about 67 millions of miles. Its apparent diameter varies between 9½″ in superior conjunction, and 62″ in inferior conjunction. The real diameter is about 7500 miles; in other words Venus is nearly as large as the Earth.

CHAPTER V.

THE EARTH.

To us, as its inhabitants, the Earth appeals in two characters, and in writing a book on astronomy it is necessary, yet difficult, to keep these two characters separate. The Earth is an ordinary planet member of the solar system, amenable to the same laws, impelled by the same forces, and going through the same movements as the other members of the Sun's *entourage*. Yet, by reason of the fact that we are ourselves on the Earth and are not spectators of it looking at it from at a distance, there are many phenomena coming under our notice which require special treatment, and it is often very difficult to say where the province of the astronomer ends and that of the geographer begins. This volume being specially designed to deal with astronomical matters, I shall pass over many subjects which may be said to be on the border line, and which some of my readers may therefore be disappointed not to find discussed. Besides the geographer, the geologist and his scientific brother the mineralogist are concerned with the Earth regarded as a planet moving through space as the other planets do. The geologist studies the actual structure of the Earth, its circumstances and history so far as they have been revealed to us, whilst the mineralogist investigates and names the materials of which it is composed, and classifies such materials with the assistance of the geologist on the one hand and of the chemist on the other. All these subordinate sciences—subordinate I mean from an astronomer's point of view—open up very varied, instructive, and interesting fields of study, but they are of course foreign to the purpose of the present volume.

Though the Earth is commonly regarded as a sphere it is not that in reality, because it is not of identical dimensions from east to west and from north to south. It is somewhat flattened at the poles; its polar diameter is less than its equatorial diameter, in the ratio of about 298 to 299, or, expressed in miles, its polar diameter is about 26 miles less than its equatorial diameter. If a globe 3 feet in diameter be taken to represent the Earth, then the polar diameter will, on this scale, be $\frac{1}{8}$ inch too long. This flattening of the poles of the Earth finds its counterpart, so far as we know, in most, and probably in all of the planets. It is most considerable and therefore most conspicuous in the case of Jupiter. It ought here to be added that a suspicion exists that the equatorial section of the Earth is not a perfect circle, but that the diameter of the Earth, taken through the points on the equator marked by the meridians 13° 58′ and 193° 58′ east of Greenwich, is one mile longer than the diameter at right angles to these two points.

The science which inquires into matters of this kind, including besides the figure of the Earth, the length of the degree at different latitudes, and the distances of places from one another, alike in angular measure and in time, is called Geodesy; it is, in point of fact, land-surveying on a very large scale, in which instruments and processes of astronomical origin are brought into operation, and in which astronomers are more or less required to take the lead.

Although we all of us now perfectly understand that the Earth is a planet moving round the Sun as a centre, it is, comparatively speaking, but recently that this fact has become generally recognised and understood. It is true that we can discover here and there in ancient writings some trace of the idea, yet it is doubtful whether 2000 years ago more than a few "advanced" thinkers thoroughly and clearly accepted it as a distinct truth. It was much more in consonance with popular thought and the actual appearance of things that the Earth should be the centre round which the Sun revolved and on which the planets depended; and accordingly, sometimes in one shape and sometimes in another, the notion of the Earth being the centre of the universe was generally accepted. The contrary opinion had, however, a few sympathisers. For instance, Aristarchus of Samos, who lived in the third century before the Christian era, supposed, if we may trust the testimony of Archimedes and Plutarch, that the Earth revolved round the Sun; this, however, was regarded as a "heresy," in respect of which he was accused of "impiety." Some few years elapsed and a certain Cleanthes of Assos is said by Plutarch to have suggested that the great phenomena of the universe might be explained by assuming that the Earth was endued with a motion of translation round the Sun together with one of rotation on its own axis. The historian states that this idea was so contrary to the received opinions that it was proposed to put Cleanthes on his trial for impiety.

In former times the philosophers who studied the solar system ranged themselves in several "schools of thought," to use a modern hackneyed phrase. Some upheld the Ptolemaic system, which took its name from a great Egyptian astronomer, Claudius Ptolemy, though it does not appear that he was actually the first to suggest it. The Ptolemaic system regarded the Earth as the centre, with the following bodies, all called planets, revolving round it in the order stated:—the Moon, Mercury, Venus, the Sun, Mars, Jupiter, and Saturn. It will be observed that there are seven bodies here named, and as seven was regarded as the "number of perfection," it was in later times considered that only these seven bodies (neither more nor less) could really be the Earth's celestial attendants. Though Ptolemy was in one sense an Egyptian, there yet prevailed amongst the Egyptians at large another theory slightly different from Ptolemy's. According to the "Egyptian theory," Mercury and Venus were regarded as satellites of the Sun, and not as primary planets appurtenant to the Earth.

After Ptolemy's era many centuries elapsed, during which the whole subject of the solar system lay practically dormant, and it continued so until the revival of learning brought new theorists upon the scene. The most important of these was Copernicus, who, in the sixteenth century, propounded a theory which eventually superseded all others, and, with slight modifications, is the one now accepted. Copernicus placed the Sun in the centre of the system, and treated it as the point around which all the primary planets revolved. So far, so good; but Copernicus went astray on the question of the orbits of the planets. He failed to realise the true character of the curves which they follow and treated these curves as "epicycles," which word may be described as representing a complicated combination of little circles which taken together form a big one. It was left to Kepler and Newton to settle all such details on a true and firm basis. But before this stage was reached a man of the highest astronomical attainments and practical experience, Tycho Brahe, made shipwreck of his reputation as an astronomer by solemnly reviving the idea of the Earth being the immovable centre of everything. He treated the Moon as revolving round the Earth at no great distance and the Sun as doing the same thing a little farther off; the five planets revolving round the sun as solar satellites. The "Tychonic system," as it is called, has something in common with the Ptolemaic system without being by any means as logical as the latter. That such far-fetched ideas as Tycho's should have been palmed off on the world of science so recently as 300 years ago is passing strange; but the explanation appears to be that his action arose out of a misconception of certain passages of Holy Scripture, which seemed irreconcilable with the Copernican theory. It must not be forgotten that Copernicus's famous book, published in 1543, in which he had announced his views, had been condemned by the Papal "Congregation of the Index;" and therefore Tycho might have had as a further motive a desire to curry favour with the authorities of the Church of Rome, and to gratify his own vanity at the same time.

With these explanations it will no longer be misleading if, for convenience sake, I speak of a certain great circle of the heavens as apparently traversed by the Sun every year, owing to the revolution of our Earth round that body. This circle is called the "Ecliptic," and its plane is usually employed by astronomers as a fixed plane of reference. It must be distinguished from that other great circle called the "celestial equator," which is the plane of the Earth's equator extended towards the stars. The plane of the equator is inclined to the ecliptic at an angle of about 23½°, which angle is known as the "obliquity of the ecliptic." It is this inclination which gives rise to the seasons which follow one another in succession during our annual journey round the Sun. The two points where the celestial equator and the ecliptic intersect are called the "equinoxes," of spring or autumn as the case may be; the points midway between these being the "solstices," of summer or winter as the case may be. These words need but little explanation, at any rate, as regards those persons who are able to trace the Latin origin of the words. "Equinox" is simply the place occupied by the Sun twice every year (namely about March 20 and September 22), when day and night are theoretically equal throughout the world, when also the sun rises exactly in the east and sets exactly in the west. The "solstices" represent the standing still of the sun at the given times and places, and are the neutral points where the Sun attains its greatest northern or southern declination. This usually occurs about June 21 and December 21. It must not be forgotten by the way, that the above application of the words "summer" and "winter" to the solstices is only correct so far as concerns places in northern terrestrial latitudes—Europe and the United States, for instance. In southern terrestrial latitudes—for instance, when speaking of what happens at the Cape of Good Hope and in Australia—the words must be reversed.

We have seen in a previous chapter that whilst the orbits of the planets are nearly true circles, none of them are quite such: and the departure from the truly circular form results in some important consequences. Whilst some of these are too technical to be explained in detail here, one at least must be referred to because of what it involves. Not only is the Earth's orbit eccentric in form, but its eccentricity varies within narrow limits; and besides this the orbit itself, as a whole, is subject to a periodical shift of place, from the joint effect of all which changes it comes about that our seasons are now of unequal length, the spring and summer quarters of the year unitedly extending to 186 days, whilst the autumn and winter quarters comprise only 178 days. The sun therefore has the chance of shining for a longer absolute period of time over the northern hemisphere than over the southern hemisphere; hence the northern is the warmer of the two hemispheres, because it has a better, because a longer, chance of storing up an accumulation of solar radiant heat. Probably it is one result of this that the north polar regions of the Earth are easier of access than the south polar regions. In the northern hemisphere navigators have reached to 81° of latitude, whereas 71° is the highest limit yet attained in the southern hemisphere. Readers who have studied the history of explorations in the Arctic regions will not need to be reminded of the controversy which has so often arisen respecting the existence or non-existence of an "Open Polar Sea."

It has already been hinted that it is not an easy matter to determine, when dealing with the Earth, where astronomy and its allied sciences, geography, geodesy and geology respectively, begin and end. But as certain topics connected with these sciences, such as the rotundity of the Earth and its rotation on its axis, will come more conveniently under consideration in other volumes of this series, I shall pass them over and only treat of a few things which more directly concern the student of nature observing either with or without the assistance of a telescope.

The fact that the Earth is surrounded by a considerable atmosphere largely composed of aqueous vapour has a material bearing on the success or failure of observations made on the Earth of bodies situated at a distance. It may be taken as a general rule that the nearer an observer is to the surface of the sea, or otherwise to the surface of the land at the sea-level, the greater will be the difficulty which will confront him in carrying on astronomical observations. Hence such observations are generally made with unsatisfactory results on the sea coast or on the banks of rivers. An interesting but rather ancient illustration of this last-named fact is to be found in the circumstance that Copernicus, who died at the age of 70, complained in his last moments that much as he had tried he had never succeeded in detecting the planet Mercury, a failure due, as Gassendi supposed, to the vapours prevailing near the horizon at the town of Thorn on the banks of the Vistula where the illustrious philosopher lived.

The phenomena depending on the presence of aqueous vapour in the atmosphere which especially come under the notice of the astronomer are Refraction, Twilight, and the Twinkling of the Stars.

Refraction is what it professes to be, a bending, and what is bent is the ray of light coming from a celestial object to a terrestrial station. Olmsted has put the matter in this way:—"We must consider that any such object always appears in the direction in which the *last* ray of light comes to the eye. If the light which comes from a star were bent into fifty directions before it reached the eye, the star would nevertheless appear in a line described by the ray nearest the eye. The operation of this principle is seen when an oar, or any stick, is thrust into the water. As the rays of light by which the oar is seen have their direction changed as they pass out of water into air, the apparent direction in which the body is seen is changed in the same degree, giving it a bent appearance—the part below the water having apparently a different direction from the part above." The direction of this refraction is determined by the general law of optics that when a ray of light passes out of a rarer into a denser medium (for instance out of air into water, or out of space into the Earth's atmosphere) it is bent *towards* a perpendicular to the surface of the medium; but when it passes out of a denser into a rarer medium it is bent *from* the perpendicular. The effect of refraction is to make a heavenly body appear to have an apparent altitude greater than its true altitude, so that, for example, an object situated actually *in* the horizon will appear *above* it. Indeed it sometimes happens that objects which are actually below the horizon and which otherwise would be invisible were it not for refraction are thus brought into sight. It was in consequence of this that on April 20, 1837, the Moon rose eclipsed before the Sun had set.

Sir Henry Holland thus alludes to the phenomenon :—" I am tempted to notice a spectacle, having a certain association with this science, which I do not remember to have seen recorded either in prose or poetry, though well meriting description in either way. This spectacle requires, however, a combination of circumstances rarely occurring —a perfectly clear Eastern and Western horizon, and an entirely level intervening surface, such as that of the sea or the African desert—the former rendering the illusion, if such it may be called, most complete to the eye. The view I seek to describe embraces the orb of the setting Sun, and that of the full Moon rising in the East—*both above the horizon at the same time.* The spectator on the sea between, if he can discard from mental vision the vessel on which he stands, and regard only these two great globes of Heaven and the sea-horizon circling unbroken around him, gains a conception through this spectacle clearer than any other conjunction can give, of those wonderful relations which it is the triumph of astronomy to disclose. All objects are excluded save the Sun, the Moon, and our own Globe between, but these objects are such in themselves that their very simplicity and paucity of number enhances the sense of the sublime. Only twice or thrice, however, have I witnessed the sight in its completeness—once on a Mediterranean voyage between Minorca and Sardinia —once in crossing the desert from Suez to Cairo, when the same full Moon showed me, a few hours later, the very different but picturesque sight of one of the annual caravans of Mecca pilgrims, with a long train of camels making their night march towards the Red Sea." *

It is due to the same cause that the Sun and the Moon when very near the horizon may often be noticed to exhibit a distorted oval outline. The fact simply is, that the upper and the lower limbs undergo a different degree of refraction. The lower limb being nearer the horizon is more affected and is consequently raised to a greater extent than the upper limb, the resulting effect being that the two limbs are seemingly squeezed closer together by the difference of the two refractions. The vertical diameter is compressed and the circular outline becomes thereby an oval outline with the lesser axis vertical and the greater axis horizontal.

Though the foregoing information merely embraces a few general principles and facts, the reader will have no difficulty in understanding that refraction exercises a very inconvenient disturbing influence on observations which relate to the exact places of celestial objects. No such observations are available for mutual comparison, however great the skill of the observer, or the perfection of his instrument, unless, and until certain corrections are applied to the observed positions in order to neutralise the disturbing effects of refraction. In practice this is usually done by means of tables of corrections, those in most general use being Bessel's. Inasmuch as refraction depends upon the aqueous vapour in the atmosphere, its amount at any given moment is affected by the height of the barometer and the

---

* "*Recollections of Past Life*" 2nd ed., p. 305.

temperature of the air. Accordingly when, for any purpose, the utmost precision is required, it is necessary to take into account the height of the barometer and the position of the mercury in the thermometer at the moment in question. At the zenith there is no refraction whatever, objects appearing projected on the background of the sky exactly in the position they would occupy were the earth altogether destitute of an atmosphere at all. The amount of the refraction increases gradually, but in accordance with a very complex law, from the zenith to the horizon. Thus the displacement due to refraction which at the zenith is nothing and at an altitude of 45° is only 57″ becomes at the horizon more than ½°. One very curious consequence is involved in the fact that the displacement due to refraction is at the horizon what it is; the diameter both of the Sun and Moon may be said to be ½°, more or less, so that when we see the lower edge of either of these luminaries just *touching* the horizon *in reality* the whole disc is completely *below* it, and would be altogether hidden by the convexity of the earth were it not for the existence of the earth's atmosphere and the consequent refraction of the rays of light passing through it from the Sun (or Moon) to the observer.

Twilight is another phenomenon associated with astronomical principles and effects which depends in some degree on the Earth's atmosphere and on the laws which regulate the reflection and refraction of light. After the Sun has set it continues to illuminate the clouds and upper strata of the air just as it may often be seen shining on the tops of hills long after it has disappeared from the view of the inhabitants of the plains below, and indeed may illuminate the chimneys of a house when it is no longer visible to a person standing in the garden below. The air and clouds thus illuminated reflect some of the Sun's light to the surface of the earth lying immediately underneath, and thus produce after sun-set and before sun-rise, in a degree more or less considerable according as the Sun is only a little or is much depressed below the horizon, that luminous glow which we call "twilight." This word is of Saxon origin and implies the presence of a twin, or double, light. As soon as the Sun has disappeared below the horizon all the clouds overhead continue for a few minutes so highly illuminated as to reflect scarcely less light than the direct light of the Sun. As, however, the Sun gradually sinks lower and lower, less and less of the visible atmosphere receives any portion of its light, and consequently less and less is reflected minute by minute to the Earth at the observer's station until at length the time comes when there is no sunlight to be reflected—and it is night. The converse of all this happens before and up to sun-rise; night ceases, twilight ensues, gradually becoming more definite; the dawn appears, and finally the full Sun bursts forth. It may here be stated as a note by the way that the circumstances under which the Sun first shows itself after it has risen above the horizon has some bearing on the probable character of the weather which is at hand. When the first indications of day-light are seen above a bank of clouds it is thought to be a sign of wind; but if the first streaks of light are discovered low down, that is in, or very near the horizon, fair weather may be expected.

Twilight is usually reckoned to last until the Sun has sunk 18° below the horizon, but the question of its duration depends on where the observer is stationed, on the season of the year, and (in a slight degree) on the condition of the atmosphere. The general rule is that the twilight is least in the tropics and increases as the observer moves away from the equator towards either pole. Whilst in the tropics a depression of 16° or 17° is sufficient to put an end to the phenomenon, in the latitude of England a depression of from 17° to 20° is required. As implied above, it varies with the latitude; and as regards the different seasons of the year, it is least on March 1 and October 12, being three weeks before the vernal equinox and three weeks after the autumnal equinox. The duration at the equator may be about 1 hour 12 minutes; it amounts to nearly 2 hours at the latitude of Greenwich, and so on towards the pole. At each pole in turn the Sun is below the horizon for 6 months, but as it is less than 18° below the horizon for about 3½ of those 6 months it may be said that there is a continual twilight for those 3½ months. Something of the same sort of thing as this occurs in the latitude of Greenwich, for there is no true night at Greenwich from May 22 to July 21, but constant twilight from sunset to sunrise, or 2 months of twilight in all. Though twilight at the equator is commonly set down as lasting about an hour, this period is there, as elsewhere, affected by the elevation of the observer above the sea-level. Where the air is very rarified, as at places situated as Quito and Lima are, the twilight is said to last no more than 20 minutes, and this would accord with the theory that where there is no air at all (*e. g.*, on the Moon) there is no twilight at all. The greater purity and clearness of mountain air, rarified as it is, is another cause which contributes to vary by reducing the duration of twilight.

It is sometimes stated that a secondary twilight may be noticed, and Sir John Herschel has spoken of it as "consequent on a re-reflection of the rays dispersed through the atmosphere in the primary one. The phenomenon seen in the clear atmosphere of the Nubian Desert, described by travellers under the name of the 'afterglow,' would seem to arise from this cause." I am not acquainted with any records which throw light on these remarks of Sir John Herschel.

The phenomenon of twinkling is a subject which has been much neglected, possibly on account of its apparent, but only apparent, simplicity. The familiar verse of our days of childhood—

"Twinkle, twinkle little star,
How I wonder what you are,
Up above the earth so high,
Like a diamond in the sky,"

contains even in this simple form a good deal of food for reflection; whilst the new version—

"Twinkle, twinkle little star,
Now we've found out what you are,
When unto the midnight sky
We the spectroscope apply,"

does so yet more.

As an optical phenomenon the twinkling, or to use the more scientific phrase, the scintillation, of the stars is a matter which has been strangely ignored by physicists. Indeed, the only investigators who seem to have dealt with it in any sort of detail are two Italians, Secchi and Respighi, Dufour, a Frenchman, Montigny, a Belgian, and the Rev. E. Ledger, an Englishman. Secchi has truly remarked that the twinkling of the stars is one of the most beautiful of the minor phenomena of the heavens. Light, sometimes bright, sometimes feeble, sometimes white, sometimes red, darts about in intermittent gleams, like the sparkling flashes of a well-cut diamond, and works upon the feelings of even the most stolid spectator. The theory of twinkling is still surrounded by many difficulties. One thing, however, is certain —it has nothing to do with recurrent changes in the intrinsic light or physical condition of the star itself, but arises during the passage of its rays through our atmosphere; it depends, therefore, in some way or other on the varying conditions of the atmosphere. On the summit of high mountains, according to the observations of all careful observers (notably Tacchini, who studied the subject on Mount Etna), the light of the stars is steady, like that of the planets; and it is so likewise during the hours of calm which often precede terrestrial storms. The vibrations are usually more frequent near the horizon, and diminish with the elevation of the star above the horizon; in other words, with the lessening of the thickness of the atmospheric strata which the rays of light have to traverse. Nevertheless, during windy weather, and specially with northerly wind, it may be noticed that the stars twinkle high up above the horizon, and even as far as the zenith. From these and other similar considerations we are justified in drawing the conclusion that twinkling largely depends on the condition and movements of the atmosphere.

Secchi further points out that it is impossible to study carefully with the naked eye all the features of twinkling, and that telescopic assistance is imperatively necessary. When, with the aid of a telescope, we scrutinise a star during a disturbed evening marked by much twinkling we see an image diffused and undefined and surrounded by rays, as if several images were superposed, and were jumping about rapidly. On such occasions we do not see that little defined disc surrounded by motionless diffraction rings, ordinarily indicative of a tranquil atmosphere. With a telescope armed with a medium power, the field of view of which is more extensive than that of a high power, we find that if a light tap is given to the telescope, the ordinary simple image is changed into a luminous curve, the perimeter of which is formed entirely of a succession of arcs exhibiting the colours of the rainbow. This coloured curve does not, in principle, differ from what one sees on swinging round and round in the air such a thing as a stick, the end of which is alight, having been freshly taken from a fire. The glowing tip produces in appearance a continuous arc, the result of the persistence of the image of the tip on the retina. In such a case the colour is constant, because the illumination resulting from the blazing wood does not vary; but in the case of a star the arcs are differently coloured during the very brief space of time in which the vibrating telescope transports the image from one side to another of the visible field. This experiment is from its nature very crude, but the idea was improved upon and reduced to a systematic shape by Montigny, who introduced into his telescope, at a certain distance from the eye-piece, a concave lens eccentrically placed with respect to the axis of the instrument, and endued with a rapid movement of rotation imparted by suitable mechanism. He thus obtained images which revolved with regularity, and so was able to submit certain features of the phenomenon to a definite system of measurement. To cut a long story short, Montigny started with the assumption (made good by the sequel) that possibly stars were affected in their twinkling by intrinsic constitutional differences; and that possibly Secchi's classification of stars into four types (a classification which depends on the spectra which they yield) might put him on the track of some intelligible conclusions with respect to the theory of twinkling.*

The results he ultimately arrived at were, that the yellow and red stars of the IInd and IIIrd types twinkle less rapidly than the white stars of the Ist type. Whilst the average number of scintillations per second of the stars of type III. were 56, those of type II. were 69, and those of type I. 86. These differences may be confidently said to depend upon too many observations of too many different stars to be fortuitous. Montigny also arrived at a number of incidental conclusions of considerable interest. The one main thread running through them, is that there is a connection between the twinkling of a star and its spectrum, which had never before been thought of. We are justified, indeed, in going so far as to

---

* For some information respecting these Secchi "Types" of Stars, see my "*Story of the Stars*," 2nd ed., p. 140.

say, that Montigny's observations point distinctly to a law on this subject, the law being that the more the spectrum of a star is interrupted by dark lines, the less frequent are its scintillations. The individual character of the light, therefore, emitted by any given star appears to affect its twinkling, both as regards the frequency thereof and the colours displayed.

Montigny collected some other interesting facts with reference to twinkling, which may here be stated in a concise form. There is a greater display of twinkling in showery weather, than when the atmosphere is in a normal condition; and in winter than in summer, whatever may be the weather. In dry weather in Spring and Autumn the twinkling is about the same, but wet has more effect in Autumn than in Spring in developing the phenomenon. Variations in the barometric pressure and in the humidity of the air also affect the amount of twinkling; there is more before a rainy period, likely to last 2 or 3 days, than before a single, or, so to speak, casual rainy day. Twinkling also varies with the aggregate total rain-fall of any group of days, being more pronounced as the rain-fall is greater, but decreasing suddenly and considerably as soon as the rainy condition of the atmosphere has passed away. The number of scintillations found to be observable with the aid of Montigny's instrument (which he called a "*scintillometre*"), varied from a minimum of 50 during June and July, to 97 in January, and 101 in February, increasing and decreasing in regular sequence from month to month. When an Aurora Borealis is visible, there is a marked increase in the amount of twinkling. It would be interesting to follow up this last named discovery by an endeavour to ascertain whether the fluctuations which are coincident in point of time with an Auroral display depend upon optical considerations connected with the Aurora, or on physical considerations having any relation to the increased development of terrestrial magnetism.

I have been thus particular in unfolding somewhat fully the present state of our knowledge concerning the twinkling of the stars, because it is evident that there are many interesting points connected with it, which may be studied by any patient and attentive star-gazer, and which do not need the instrumental appliances and technical refinements which are only to be found in fully-equipped public and private observatories.

It should be mentioned in conclusion that the planets twinkle very little, or, more often, not at all. This is mainly due to the fact that they exhibit discs of sensible diameter and therefore that there is, as Young puts it, "a general unchanging average of brightness for the sum total of all the luminous points of which the disc is composed. When, for instance, point A of the disc becomes dark for a moment, point B, very near to it, is just as likely to become bright; the interference conditions being different for the 2 points. The different points of the disc *do not keep step*, so to speak, in their twinkling." The non-twinkling of planets because they possess sensible discs is often available as a means for determining when a planet is looked for, which, of several objects looked at, is the planet wanted and which are merely stars.

CHAPTER VI.

THE MOON.

THE Moon being merely the satellite of a planet, to wit, the Earth, it should, according to the plan of this book, be included in the chapter which deals with its primary; but for us inhabitants of the Earth the Moon has so many special features of interest that it will be better to give it a special chapter to itself.

We may regard the Moon in a twofold aspect, and consider what it is as a mere object to look at, and what it does for us; probably my present readers will prefer that most prominence shall be given to the former aspect. The Moon as seen with the naked eye exhibits a silvery mass of light, which at the epoch of what is called "full Moon" has a seemingly even circular outline. Full or not full, its surface appears to be irregularly shaded or mottled. The immediate cause of this shading is the fact that the surface of the Moon, not being really smooth, reflects irregularly the Sun's light which falls upon it. The *causa causans* of this is the existence of numerous mountains and valleys on its surface, and which were first discovered to be such by Galileo. That there are mountains is proved by the shadows cast by their peaks on the surrounding plains, when the Sun illuminates the Moon obliquely—that is, when the Moon is shining either as a crescent or gibbous. Such shadows, however, disappear at the phase of "Full-Moon," because the Sun's rays then fall perpendicularly on the Moon's surface. When the Moon presents either a crescent or a gibbous form (in point of fact when it presents any form except that of "Full-Moon"), the boundary line which separates the illuminated from the unilluminated portion (and which boundary line is generally spoken of as the "terminator") has a rough, jagged appearance; this is due to the fact that the Sun's light falls first on the summits of the peaks, and that the adjacent valleys and declivities are in shade. These remain so till by reason of the Moon's progress in its orbit a sufficient time has elapsed for the Sun to penetrate to the bottom of the valleys. With this explanation the reader will have no difficulty in realising why the terminator always exhibits an irregular or jagged edge.

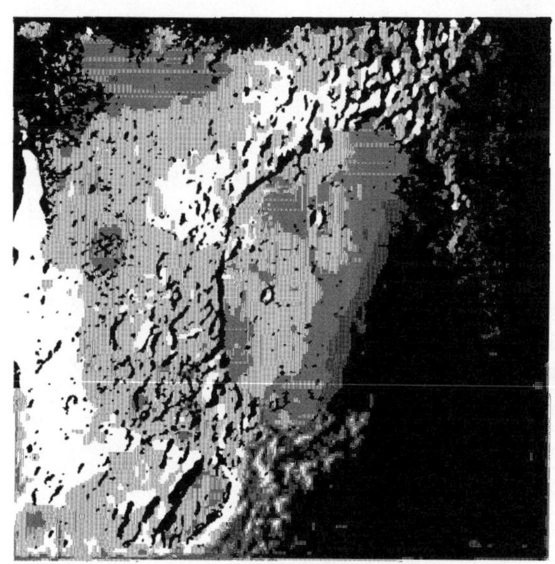

FIG. 11.—Mare Crisium.
(Lick Observatory photographs.)

Various mountains on the Moon to the number of more than a thousand have been mapped, and their elevations calculated. Of these fully half have received names, being those of men of various dates and nationalities, who have figured conspicuously in the annals of science, including some, however, who have not done so. Whilst many of these mountains are isolated elevations, not a few form definite chains of mountains, and to certain of these chains definite names, borrowed from the Earth, have been given. Thus we find on maps of the Moon the "Apennines," the "Alps," the "Altai Mountains," the "Dorfel Mountains," the "Caucasus Mountains," and so on.

Besides the mountains there exist on the Moon a number of plains analogous in some sense to the "steppes" of Asia and the "prairies" of North America. These were termed "seas" in the early days of the telescope, because it was assumed that as they were so large and so smooth they were vast tracts of water. This supposition has long ago been overthrown, but the names have been retained as a matter of convenience. Hence it comes about that in descriptions of the Moon one meets with such names as *Mare Imbrium*, the "Sea of Showers"; *Mare Serenitatis*, the "Sea of Serenity"; *Mare Tranquillitatis*, the "Sea of Tranquillity"; and so on. It seems probable that the so-called seas represent in nearly its original form what was once the original surface of the Moon before the mountains were formed. A confirmation of this idea is to be found in the fact that though these plains are fairly level surfaces compared with the masses of mountains which hedge them in on all sides, yet the plains themselves are dotted over with inequalities (small elevations and pits), which seem to suggest that some of them might eventually have developed into mountains if the further formation of mountains had not been arrested by the fiat of the Creator.

Though hitherto we have been speaking of the mountains of the Moon under that generic title, it is necessary for the reader to understand that the Moon's surface exhibits everywhere remarkable illustrations of those geological processes which we on the earth associate with the word "volcano." There cannot be the least doubt that the existing surface of the Moon, as we see it, owes all its striking features to volcanic action, differing little from the volcanic action to which we are accustomed on the earth. That this theory is well founded may be very easily inferred by comparing the structural details of certain terrestrial volcanoes and their surroundings with a typical lunar mountain, or indeed, I might say, with any lunar mountain. This point was very well worked out some 40 years ago by Professor Piazzi Smyth, who placed on pictorial record his results of an examination and survey of the Peak of Teneriffe. Any person seeing side by side one of Smyth's pictures of Teneriffe and a picture of any average lunar crater would find great difficulty if the pictures were not labelled in determining which was which.

The one special feature of the Moon, which never fails to attract the attention of everybody who looks at our satellite for the first time through a telescope, are the crater mountains, which indeed constitute an immense majority of all the lunar mountains. Their outline almost always conforms, more or less, to that of the circle, but when seen near either limb of the Moon they often appear considerably oval simply because they are then seen considerably foreshortened. In their normal form they exhibit a basin bounded by a ridge, with a conical elevation in the centre of the basin, the basin and the cone together being evidently the result of an uprush of gases breaking through the outer crust of the Moon and carrying with them masses of molten lava. This lava, with perhaps the materials in fragments, projected in the first instance up into the air, fell back on to the Moon forming first of all the outer edge of the basin, and subsequently, as the eruptive force became weakened, the small central accumulation, which took, as it naturally would do, a conical shape. An experimental imitation of the process thus inferred was carried out some years ago by a French physicist, Bergeron, who acted upon a very fusible mixture of metals known as Wood's alloy by forcing through it a current of hot air. The success of this experiment was complete, and Bergeron considered that his experiments, taken as a whole, were calculated to throw much light on the past history of the Moon.

Several observers at various times have fancied they have seen signs that the lunar mountain Aristarchus was an active volcano even up to the present century; but it admits of no doubt that this idea is altogether a misconception, and that what they saw as a faint illumination of the summit of Aristarchus was no more than an effect of earth-shine. On the general question of volcanic action on the Moon, Sir John Herschel summed up as follows:—"Decisive marks of volcanic stratification arising from successive deposits of ejected matter, and evident indications of lava currents, streaming outwards in all directions, may be clearly traced with powerful telescopes. In Lord Rosse's magnificent Reflector the flat bottom of the crater called Albategnius is seen to be strewed with blocks not visible in inferior telescopes, while the exterior ridge of another (Aristillus) is all hatched over with deep gulleys radiating towards its centre."

The valleys and clefts or rills visible on the Moon's surface constitute another remarkable feature in the topography of our satellite. The valleys, properly so-called, require no particular comment, because they are just what their name implies—hollows often many miles long and several miles wide. The clefts or rills, however, are more mysterious, by reason of their great length and remarkable narrowness. One is almost led to infer that they are naught else but cracks in the lunar crust, the result of sudden cooling, how caused is of course not known.

There is another lunar feature to be mentioned somewhat akin to the foregoing in appearance but apparently, however, owing its origin to a different cause. I refer to the systems of bright streaks which, especially at or near the time of full Moon, are seen to radiate from several of the largest craters, and in particular from Tycho, Copernicus, Kepler and Aristarchus. These bright streaks extend in many cases far beyond what may fairly be considered as the neighbourhood of the craters from which they start, traversing distant mountains, valleys and other craters in a way which renders it very difficult to assign an explanation of their origin.

There are 13 areas on the Moon, which used to be regarded as "seas," one of them, however, bearing the name of "*Oceanus Procellarum*" the "Ocean of Storms"; but besides these there are several bays, termed in Latin *Sinus*, of which the most important is the *Sinus Iridum* or the "Bay of Rainbows," a beautiful spot on the northern border of the *Mare Imbrium*, and best seen when the Moon is between 9 and 10 days old. The summits of the semi-circular range of rocks which enclose the bay are then strongly illuminated and a greenish shadow marks the valley at its base. By the way, it is worth mentioning that not a few of the lunar seas, so-called, seem to be pervaded by a greenish hue, though no particular explanation of this fact is forthcoming.

Much controversy has ranged round the question whether or not the Moon has an atmosphere. Without doubt the preponderance of opinion is on the negative side. though it must be admitted that some observers of eminence have suggested that there are indeed traces of an atmosphere to be had, but that it is extremely attenuated and of no great extent, otherwise it must render its presence discoverable by optical phenomena which it is certain cannot be detected.

A brief reference may here be made to a curious phenomenon sometimes seen in connection with occultations of stars by the Moon. Premising that an "occultation" is the disappearance of a star behind the solid body of the Moon by reason of the forward movement of the Moon in her orbit, it must be stated that though generally the Moon extinguishes the star's light instantaneously, yet this does not invariably happen, for sometimes the star seems to hang upon the Moon's limb as if reluctant to disappear. No very clear or satisfactory explanation of this phenomenon has yet been given; the existence of a lunar atmosphere would be an explanation, and accordingly this anomalous appearance, seen on occasions, has been advanced in support of the theory that a lunar atmosphere does exist; but, nevertheless, astronomers do not accept that idea.

Any one desirous of carrying out a careful

study of the Moon's surface must be provided with a good map, and for general purposes none is so convenient or accessible as Webb's, reduced from Beer and Madler's *Mappa Selenographica* published in 1837, of which another reproduction is given in Lardner's *Astronomy*. Those, however, who would desire to study the Moon with the utmost attention to detail must provide themselves with Schmidt's map published in 1878 at the expense of the German Government. When it is stated that this map represents the Moon on a circle $7\frac{1}{2}$ feet in diameter, the size and amount of detail in it will be readily understood. Special books on the Moon furnishing numerous engravings and detailed descriptions have been written by Carpenter and Nasmyth (jointly) and by Neison.

Various attempts have been made to determine the amount of light reflected by the Moon, and also the question whether it yields any measurable amount of heat. As regards the light of the full Moon compared with that of the Sun, the estimates range from $\frac{1}{300000}$ to $\frac{1}{800000}$, a discrepancy not perhaps greater than might be expected under the circumstances of the case.

With respect to the heat possessed by, or radiated from the Moon's surface, the conclusions of those who have attempted to deal with the matter are less consistent. As regards the surface of the Moon itself Sir John Herschel was of opinion that it is heated at least to the temperature of boiling water, but that owing to the radiant heat having to pass through our atmosphere, which acts as an obstacle, it is no wonder that it should be difficult for us to become conscious of its existence. In 1846 Melloni, by concentrating the rays of the Moon with a lens 3 feet in diameter, thought he detected a sensible elevation of temperature; and in 1856 C. P. Smyth at Teneriffe, but with inferior instrumental appliances, arrived at the same conclusion. Though Professor Tyndall in 1861 obtained a contrary result, yet the most recent experiments by the younger Earl of Rosse, Professor Langley, and others, all tend to show that the Moon does really radiate a certain infinitesimally small amount of heat. Perhaps, however, it will be best to give Langley's ideas as to this in his own words:—"While we have found abundant evidence of heat from the Moon, every method we have tried, or that has been tried by others, for determining the character of this heat appears to us inconclusive; and without questioning that the Moon radiates heat earthward from its soil, we have not yet found any experimental means of discriminating with such certainty between this and reflected heat that it is not open to misinterpretation." It is obvious from the foregoing that we on the Earth need not concern ourselves very much about lunar heat; and I will only add that F. W. Very, by an ingenious endeavour to localise the Moon's radiant heat, has been able, he thinks, to establish the fact that on the part of the Moon to which the Sun is setting, what he calls the heat-gradient (using a phrase suggested by terrestrial meteorology) appears to be steeper than on that part to which the Sun is rising. Generally, Very's observations accord fairly with Lord Rosse's.

The Moon revolves round the Earth in 27 d. 7 h. 43 m. 11 s. at a mean distance of 237,300 miles, in an orbit which is somewhat, but not very, eccentric. Its angular diameter at mean distance is 31′ 5″, or, say, just over ½°. The real diameter may be called 2160 miles.

A few words will probably be expected by the reader on the subject of lunar influences on the weather, and generally; this being a matter highly attractive to the popular mind. The truth appears to lie, as usual, between two extremes of thought. The Moon, of course, is the main cause of the tides of the Ocean, and it is not entirely inconceivable that tidal changes imparted to vast masses of water may be either synchronous with, or may in some way engender, analogous movements in the Earth's atmosphere; though no distinct proofs of this, as a determinate fact, can be brought forward.

There is no doubt whatever that at or near the time of full Moon, evening clouds tend to disperse as the Moon comes up to the meridian, and that by the time the Moon has reached the meridian a sky previously overcast will have become almost or quite clear. Sir John Herschel has alluded to this by speaking of a "tendency to disappearance of clouds under a full Moon"; and he considers this "fully entitled to rank as a meteorological fact." He goes on, not unnaturally, to suggest the obvious thought that such dissipation of terrestrial clouds is due to the circumstance that, assuming heat really comes by radiation from the Moon (and we have seen on a previous page the probability of this) such radiant heat will be more potential if it falls on the Earth perpendicularly, as from a Meridian Moon, than if it comes to us at any one locality from a Moon low down in the observer's horizon, and therefore has to pass through the denser strata of the Earth's atmosphere and suffer material enfeeblement accordingly. I am aware that Mr. Ellis, late of the Royal Observatory, Greenwich, has sought to show by a seemingly powerful array of statistics that the idea now under consideration is unfounded, but I consider that we have here only one more illustration of the familiar statement that you can prove anything you like by statistics. I am firmly convinced, as the result of more than 30 years' observation, that terrestrial clouds do disperse under the circumstances stated. Sir J. Herschel added that his statement proceeded from his own observation "made quite independently of any knowledge of such a tendency having been observed by others. Humboldt, however, in his *Personal Narrative*, speaks of it as well known to the pilots and seamen of Spanish America." Sir John Herschel further remarked:—"Arago has shown from a comparison of rain, registered as having fallen during a long period, that a slight preponderance in respect of quantity falls near the 'new' Moon over that which falls near the 'full.' This would be a natural and necessary consequence of a preponderance of a cloudless sky about the 'full,' and forms, therefore, part and parcel of the same meteorological fact."

Bernadin has asserted it to be a fact that many thunderstorms occur about the period of "new" or "full" Moon. But what I want most to warn the reader against is that popular idea (wonderfully wide-spread it must be admitted) that at the epochs of what are called, most illogically, the Moon's "changes," changes of weather may certainly be expected. There is absolutely no foundation whatever for this, and still more void of authority (if such a phrase is admissible) is a table of imaginary weather to be expected at changes of the Moon, often met with in books published half a century ago, and still occasionally reprinted in third-rate almanacs, and designated "Dr. Herschel's Weather Table." This precious production is not only devoid of authenticity as regards its name, but may easily be seen to be fraudulent in its reputed facts any month in the year.

It would be beyond both my present available space and the legitimate objects of this work to attempt even an outline of the influences over things terrestrial ascribed to, or associated, rightly or wrongly, with the Moon, and of which the word "lunatic" perhaps affords the most familiar exponent.

CHAPTER VII.

MARS.

MARS, though considerably smaller than the Earth, is commonly regarded as the planet which, taken all in all, .bears most resemblance to the Earth, though only one-fourth its size. Under circumstances which have already been briefly alluded to in Chapter I., Mars exhibits from time to time a slight phase, but nothing approaching in amount the phases presented by the two inferior planets, Mercury and Venus. When in opposition to the Sun, that is to say when on the meridian at midnight, it has a truly circular disc; but between opposition and its two positions of quadrature it is gibbous. At the minimum phase, which is at each quadrature, E. or W. as the case may be, the planet resembles the Moon 3 days from its "full." These phases are an indication that Mars shines by the reflected light of the Sun.

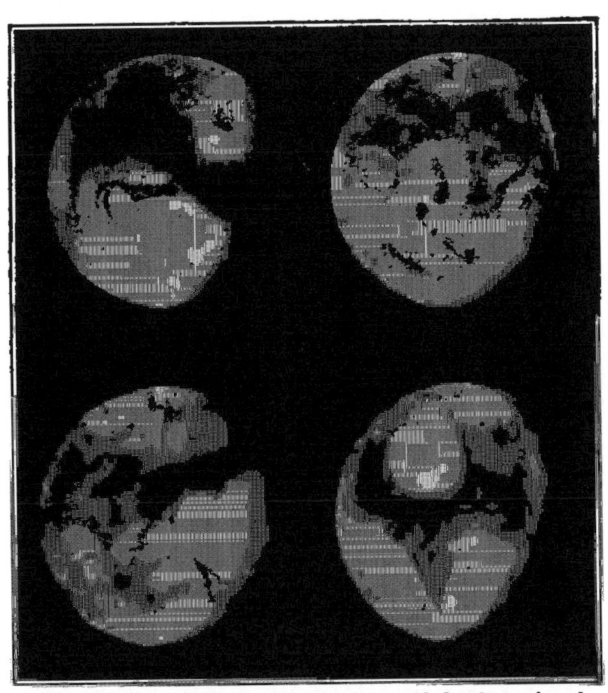

Fig. 12.—Four views of Mars differing 90° in longitude (Barnard).

It is a remarkable tribute to Galileo's powers of observation that with his trumpery telescope, only a few inches long, he should have been able to suspect the existence of a Martial phase. Writing to a friend in 1610 he says:—"I dare not affirm that I can observe the phases of Mars; however, if I mistake not, I think I already perceive that he is not perfectly round."

The period in which Mars performs its journey round the Sun (called the sidereal period) is about 687 days; but owing to the Earth's motion we are more concerned with what is called the planet's synodical period of 780 days than with its sidereal period of 687 days. The synodical period is the interval between two successive conjunctions or oppositions of the planet as regards the Earth, and 780 days being twice 365 and 50 days over, it follows that we have an opportunity of seeing the planet at its best about every 2 years; and this is one of the reasons why Mars has been so much and so thoroughly studied as regards its physical appearance. Of course Mars is not equally well seen every 2 years, because it may so happen at a given opposition that it may be at its nearest to the Sun (perihelion), and the Earth at its farthest from the Sun (aphelion), in which case the actual distance between the two bodies will be the greatest possible. What is therefore wanted is for the planet to be nearest to the Sun and nearest to the Earth at the same time, under which circumstances it shines with a brilliancy rivalling Jupiter. This favourable combination occurs once in 7 synodical revolutions, or about every 15 years. The most favourable oppositions occur at the end of August, and the least favourable at the end of February. The next very favourable opposition will not occur until 1909. Mars may approach to within about 35 millions of miles from the Earth at a favourable opposition, whilst under extreme circumstances the other way it may be no nearer than 61 millions of miles at opposition.

Mars in opposition is a very conspicuous object in the Heavens, shining with a fiery red light which has always been regarded as a peculiar attribute of the planet, so much so that its name, or epithet, in many languages conveys the idea of "fiery" or "blazing." It is recorded that in

August 1719 its brilliancy was such as to cause a panic amongst the public.

Telescopically examined, Mars is always found to exhibit patches of shade of various sizes and shapes, and, on the whole, fairly permanent from year to year. During the last few years in particular these markings have been subjected to very careful scrutiny and measurement at the hands of numerous observers of skill and experience, and armed in many cases with very powerful telescopes. The conjoint effect of the observations obtained has been largely to augment our knowledge of the planet's geography, or (to use the proper term) "areography." Before describing the minutest details recorded and pencilled by the best observers, it will be best to speak of the leading general features which are within the grasp of comparatively small telescopes —say, refractors of 6 inches and reflectors of 12 inches in aperture. The first thing which presents itself as very obvious on the disc of Mars, is the fact that certain portions are ruddy, whilst others are greenish in hue. It is generally assumed that the red areas represent land and the green areas water. On this subject Sir John Herschel's remarks, penned about half a century ago, may be said still to stand good. He ascribes the ruddy colour to "an ochrey tinge in the general soil, like what the red sandstone districts on the Earth may possibly offer to the inhabitants of Mars, only more decided." The propriety of this thought will be best appreciated by a reader who has travelled through parts of North Gloucestershire, and seen a succession of ploughed fields in that locality. The deep red colour of the soil is in many places very conspicuous. It has often been remarked that the redness of Mars is much more noticeable with the naked eye than with a telescope; and Arago carried this idea one step further in suggesting that the higher the optical power the less the colour. This, however, might naturally be expected.

The most prominent surface marking on Mars is that known as the "Kaiser Sea," sometimes called the "V-mark" from its resemblance to that letter, though a leg of mutton would be quite as good a simile. East of the Kaiser Sea and a little north of the planet's equator is a well-defined dark streak known as "Herschel II. Strait"; whilst on the west side is another shaded area which has been called "Flammarion Sea." These three features are so very conspicuous, that, provided the hemisphere in which they are situated is fairly in front of the observer, his telescope, if it will show anything on Mars, will show these. The white patches seen on certain occasions *at* Mars's N. pole and *close to* its S. pole form another special feature of interest connected with this planet. It admits of no doubt whatever that these are immense masses of snow and ice which undergo at stated intervals changes analogous to the changes which we know happen in the great fields of ice situated in the regions of the Earth surrounding the Earth's two poles. Not only do these white patches look like snow, but if attention is paid to the changes they undergo and the epochs at which the changes take place there will be found abundant confirmation of this theory, for these patches decrease in size when brought under the Sun's influence on the approach of summer and increase again in size when the summer is over and winter draws near. In the second half of 1892 the Southern Pole was in full view, and during especially July and August the diminution of the snow area from week to week was very evident. Schiaparelli, who observed it with great attention during that season, noted at the commencement of the season that the snow reached at the first as far as latitude 70° and formed a polar cap some 1200 miles in diameter. Its subsequent decrease, however, was so marked that two or three months later the diameter of the snow patch had dwindled to no more than 180 miles, and became indeed still smaller at a later period. The summer solstice on Mars occurred on October 13, 1892, which was therefore the epoch of midsummer for Mars's southern hemisphere. Whilst these changes were taking place in the southern hemisphere, no doubt changes of the reverse character were going on in the northern hemisphere, but they were not visible from the Earth because the North Pole was situated in that hemisphere of Mars which was turned away from the Earth. In previous years, however, the North Pole being turned towards the Earth its snow was also seen to undergo the same sort of change; in other words, was seen to melt. This happened, and was seen in 1882, 1884, and 1886. These observations of the alternate increase and decrease of the polar snow on Mars may be viewed with telescopes of moderate power, but of course it is more interesting and profitable to watch them with a large telescope. The fact (for it is an undoubted fact) that the north polar snow is concentric with the planet's axis whilst the southern polar patch is eccentric to the extent of about 180 miles from the southern pole is one which has not yet received a satisfactory explanation. If both patches were eccentric so as to be exactly opposite to one another an explanation would be much more easy for we might say that the poles of rotation lay in one direction and the poles of cold in another.

I have spoken on a previous page of three specially conspicuous shadings of Mars, and other similar shadings to the number perhaps of a couple of dozen were generally recognised by astronomers (having been mapped and named) down to about the year 1877. In that year the astronomical world was startled by the announcement that Schiaparelli of Milan, an able and competent observer, had discovered that those shaded areas which all previous astronomers had regarded as continents or vast tracts of land, were in reality islands, that is to say, so far, that the continents in question were cut up by innumerable channels intersecting one another at various angles. When this discovery was announced, and older observations and drawings came to be examined, it was found, or at anyrate thought, that these so-called canals might be traced in drawings of earlier dates by Dawes, Secchi, and Holden. So much for 1877. In December, 1881, the planet was again in opposition, but farther off in distance, and therefore smaller in size than in 1877. It was, however, higher up in the Heavens as seen at Milan and the weather appears to have been more favourable. In these altered circumstances Schiaparelli again saw his canals, but this time they were in at least as many as twenty instances seen in duplicate ; that is to say, a twin canal was seen to run parallel to the original one at a distance of from 200 to 400 miles, as the case might be. The existence of not only single canals but of twin canals seems an established fact, for Schiaparelli's drawings and descriptions have been confirmed by compe-

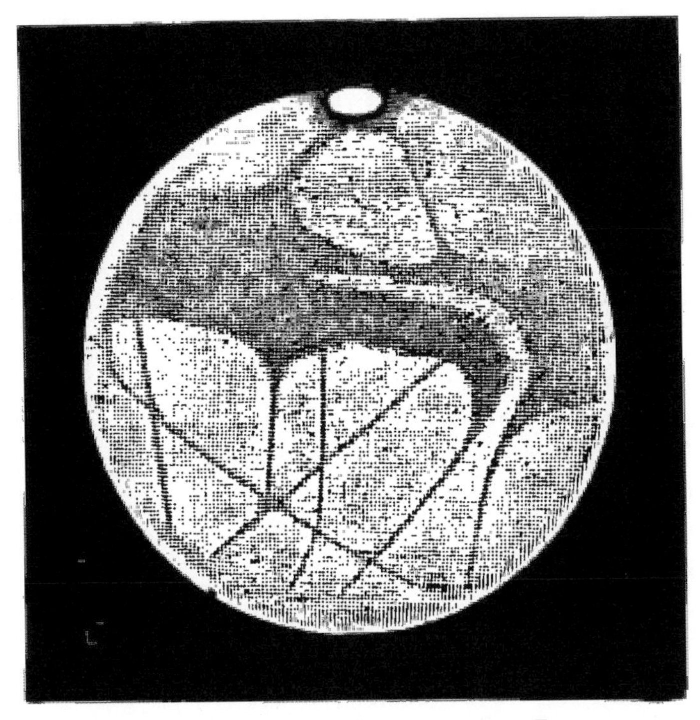

FIG. 13.—Mars, August 27, 1892 (Guiot).

tent testimony; but explanation is nowhere; especially in view of Schiaparelli's own idea that the duplication of his canals is perhaps not a permanent feature but a periodical phenomenon depending on, or connected in some way with, Mars's seasons.

Several points stand out clearly established by the observations of Mars during the opposition of 1894, especially the correctness of Schiaparelli's discoveries and maps. Most of the canals originally seen by him were again seen, and thus their existence was confirmed, whilst new ones were also noticed. Many of these canals were double.

The great extent of the S. Polar cap and its rapid disappearance as Mars's summer approached was also a special feature of the observations of 1894. It dwindled until it became almost invisible, or at best showed as a tiny speck. It is thought by some observers that as the Polar cap melts, the water collects round the Pole, and thence flows over the planet's surface, giving rise to the phenomenon of canals, and that this is the way the planet's surface is irrigated. It may here be remarked that the word "canal," which has been given to these dark streaks crossing and cutting up the large areas of land in Mars, is an unfortunate one, suggesting as it does artificial agency. But these Martial canals are probably, especially the largest, a great many miles in width and hundreds of miles in length, though some are smaller; and they are probably nature's method of distributing over the continents and lands of Mars the water which collects round the Pole during the rapid melting of the Polar snows.

The idea of the presence of cloud or mist on Mars also received strong confirmation in 1894. Large portions of the planet's disc were found to be hidden from view. "Herschel I. Continent" and the "Maraldi Sea" (both well-known markings on Mars, readily visible with small telescopes) were at times quite obscured by cloud. Indeed, the Maraldi Sea was occasionally quite blotted out: other well-known markings were also either blotted out or only faintly seen. These facts seem almost to prove conclusively the existence of cloud and vapour in Mars, especially as some of these markings subsequently again assumed their ordinary form and colour. Bright projections too were seen at times on the terminator of

Mars, giving rise to the belief that there are high mountains on the planet, though some observers regarded these projections as high clouds powerfully reflecting the Sun's light.

Mars rotates on its axis in 24h. 37m. 22s., a period so nearly coincident with the period of the Earth's rotation as greatly to facilitate the mapping of Mars's features by work continued from day to day by observers who have the necessary instrumental means and artistic skill in handling the pencil.

Mars has an atmosphere which may be said to be no more than moderately dense; that is to say much less dense than the Earth's atmosphere. Of course the existence of snow, which has been taken for granted on a previous page, carries with it the existence of water and aqueous vapour—a fact capable of independent spectroscopic proof.

The inclination of Mars's axis to the ecliptic has not been ascertained with all desirable certainty, but if Sir W. Herschel's estimate that the obliquity on Mars is 28¾° (the Earth's obliquity being 23½°) is correct, it is evident that there must be a very close similarity between the seasons of the Earth and the seasons of Mars, thereby furnishing another link of proof to support the statement made at the commencement of this chapter that, taken all in all, Mars is the planet which bears most resemblance to the Earth.

The apparent absence of satellites in the case of Mars was long a matter of regret to astronomers; they seemed to think that such a planet ought to have at least one companion. At last, in 1887, two were found by Hall at Washington, U. S., using a very fine refractor of 26 inches aperture. These satellites, which have been named Phobos and Deimos, are, however, very small, for Phobos at its best only resembles a star of mag. 11½, whilst Deimos is no brighter than a star of mag. 13½; from this it will be understood that only very large telescopes will show either of them. Phobos revolves round Mars in 7½ hours at a distance of about 6000 miles, whilst Deimos revolves in 30 hours at a distance of about 15,000 miles. It has been thought that neither of them can be more than about 6 or 7 miles in diameter, and therefore that they can not afford much light to their primary.

Mars revolves round the Sun in 686d. 23h. 30m., at a mean distance of 141 million of miles, which the eccentricity of its orbit may increase to 154 millions or diminish to 128 millions. The planet's apparent diameter varies between 4″ in conjunction and 30″ in opposition. Owing to the great eccentricity of the orbit the planet's apparent diameter as seen from the Earth varies very much at different oppositions. The real diameter is rather more than 4000 miles.

CHAPTER VIII.

THE MINOR PLANETS.

In 1772 a German astronomer named Bode, of Berlin, drew attention to certain curious numerical relations subsisting between the distances of the various planets. This "law," as it has been sometimes called, usually bears Bode's name, though it was not he but J. D. Titius of Wittemberg who really first discovered it.

Take the numbers—

0, 3, 6, 12, 24, 48, 96, 192, 384;

each of which (the second excepted) is double the preceding; adding to each of these numbers 4 we obtain—

4, 7, 10, 16, 28, 52, 100, 196, 388;

which numbers approximately represent the distances of the planets from the sun expressed in radii of the Earth's orbit. A little table will make the matter more clear.

Planets.	Distance: Bode's Law.	True distance from Sun.
Mercury	4	3·9
Venus	7	7·2
Earth	10	10.0
Mars	16	15·2
[Ceres]	[28]	[27·7]
Jupiter	52	52.0
Saturn	100	95·4
[Uranus]	[196]	[191.8]
[Neptune]	[388]	[300.0]

Bode having examined these relations and noticing the void between 16 and 52 (Ceres and the other minor planets, and Uranus and Neptune also, being then unknown) ventured to predict the discovery of new planets, and this idea stimulated him to organise a little company of astronomers to hunt for new planets. Before, however, this scheme was got into working order, Piazzi, director of the Observatory at Palermo, on January 1, 1801, noted an 8th magnitude star in Taurus, which on the next and succeeding nights he saw again, and found had moved. He observed the strange object for 6 weeks, when illness interrupted him. However he wrote letters announcing what he had seen, one of them to Bode himself; but this letter, though dated Jan. 24, did not reach Bode at Berlin, till March 20—a striking illustration of the state of the Postal service on the Continent less than 100 years ago. The new body, at first assumed to be a tailless comet, was eventually recognised to be a new planet; and the name of Ceres, the tutelary goddess of Sicily, was at Piazzi's instance bestowed upon it.

Looking for Ceres in March, 1802, Olbers at Bremen, came upon another new planet, which was afterwards named Pallas. At first he thought he had got hold of a new variable star, but two hours sufficed to show that the object under notice was in motion. The two new bodies were found to be so much alike in size and appearance, and in their orbits, that Olbers suggested both were but fragments of some larger body which had been shattered by some great convulsion of nature. The idea was a daring one, and it was an attractive one, though now regarded as untenable. However it served the purpose of stimulating research, and the discovery of Pallas was followed by that of Juno, by Harding, at Lilienthal 1804; and of Vesta, by Olbers, at Bremen in 1807.

The organised search for minor planets was relinquished in 1816, presumably because no more planets seemed to be forthcoming, and it does not appear that any further attempts were made by anybody till about 1830, when a Prussian amateur, named Hencke of Driessen, profiting by the publication of some new star maps put forth by the Berlin Academy, commenced a methodical search for small planets. These Berlin maps, one for each hour of R. A., were only completed in 1859, and, therefore, Hencke had only a small number of them at his command during the early years of his labours. Still it is strange that 15 years elapsed before his zeal and perseverance were rewarded, his first discovery, the planet Astræa, not taking place till December 1845. Once however the ice was broken new planets followed with considerable rapidity, and beginning with 1847, no year has elapsed without several or many having been found. During the last decade the number detected annually has been very great—sometimes as many as 20 in a year, but this has been the result of photography being brought to bear on the work. It is obvious that if a photograph of a given field taken on any one day is compared with a photograph taken a few days earlier or later, and any of the objects photographed have moved, their change of place will soon be noticed and will be a distinct proof of their planetary nature.

It seems quite certain that all the larger of these planets have now been found, for the average brilliancy (and this no doubt means the average size) of those recently discovered has been steadily diminishing year by year, and it looks as if the limit of visibility will soon be reached, if it has not been reached already.

The three largest of these bodies, in order of size, have generally been thought to be Vesta, Ceres, and Pallas; but Barnard, from observations made in 1894, concluded that Ceres is 520 miles in diameter; Pallas, 304 miles; and Vesta, 241 miles. As to all the rest of the minor planets, excepting Juno, Hornstein is of opinion that those having a greater diameter than 25 geographical miles are few in number, and that the majority of them are no larger than from 5 to 15 miles in diameter.

From what has gone before the reader will readily infer that these minor planets are of no sort of interest to the casual amateur who dabbles in Astronomy; and indeed that they are of very little interest to anybody. With a few general statistics, therefore, this chapter may be concluded. The total number of minor planets now known nearly reaches 500, and every year increases the list; but not, however, at as rapid a rate as was once the case, because the German mathematicians, who alone latterly have been willing to trouble themselves with the computation of the orbits, are understood to have announced that they are no longer able to keep pace with the discoveries made. Those who care to investigate in detail the circumstances of these planets will find great extremes in the nature of the orbits. Whilst the planet nearest to the Sun has a period of only 3 years, the most distant occupies nearly 9 years in performing its journey round the Sun. So, also, there are great differences in the eccentricities of the orbits and in their inclinations to the ecliptic. Whilst one planet revolves almost in the plane of the ecliptic, another (Pallas) has an orbit which is inclined no less than 34° to the ecliptic. One word, in conclusion, as to the names applied to these bodies. At the outset the names given were, without exception, chosen from the mythologies of ancient Greece and Rome, but, latterly, the most fantastic and ridiculous names have in many cases been selected, names which in too many instances have served no other purpose than that of displaying the national or personal vanity of the astronomers who applied them to the several planets. The French are great offenders in this matter.

CHAPTER IX.

JUPITER.

THE planet Jupiter occupies, in one sense, the first position in the planetary world, it being the largest of all the planets Moreover, with the exception of Venus, it is the brightest of the planets. As with Mars, and for the like reason, Jupiter, when in the positions known as the Quadratures (or near thereto), exhibits a slight phase, but owing to the far greater distance from the Sun of Jupiter, compared with Mars, the deviation of the illuminated surface from that of a complete circle is very small; it is, however, perceptible at or near the time of quadrature, a slight shading off of the limb farthest from the Sun being traceable.

Jupiter is noteworthy on account of two features, both of them more or less familiar, at least by name, to most people—its belts and its satellites,—both of which will be described in due course.

The belts are dusky streaks, which vary from time to time both in breadth and number: most commonly two broad belts will be seen with two or three narrower ones on either side; but sometimes all are rather narrow, and their narrowness is made up for by an increase in their number.

Under all circumstances they lie practically parallel, or nearly so, to the planet's equator. It is generally thought that the planet, whatever may be its actual structure or constitution, is surrounded by a dense cloudy envelope, and that the

FIG. 14.—Jupiter, November 27, 1857 (Dawes).

shaded streaks which we call belts are rifts in this atmosphere, which expose to view the solid body of the planet underneath. Whether, however, the term "solid body" is an accurate one to be used in this connection is thought by some to be open to doubt. The laws which regulate the existence of these belts are quite unknown; indeed it seems doubtful whether any laws exist at all, for the belts at one time appear to undergo constant change, whilst at another time they remain almost unchanged for several months. It has been suggested that when the changes are rapid it must be presumed that great atmospheric storms are to be considered as in progress, and possibly this may be the true explanation. Belts are commonly non-existent immediately under the equator; whilst north and south of this void space it most usually happens that there is one broad belt and several narrower ones in each hemisphere. At each pole the planet's brightness is less than the average brightness, but it cannot exactly be said that this is due to the existence there of belts properly so called.

It was formerly considered that no tinges of colour could be traced on Jupiter except a silvery gray of different degrees of intensity; but during the last thirty years there can be no doubt that shades of brown, red, and orange, of no great depth, but yet quite definite have been traceable. Many observers concur in this opinion. Whether this detection of colour is due to an absolute development of colour during the period in question; or whether its detection is merely the result of more careful scrutiny with better instruments is a matter as to which the evidence is not clear. Though the general position of the belts is such that they are parallel to the planet's equator, yet there are sometimes exceptions to this rule, for in a few very rare instances a streak in the nature of a narrow belt has been seen, inclined to the equator at a decided angle, perhaps 20° or even more.

It occasionally happens that spots are seen on Jupiter's belts. Sometimes these remain visible for a considerable period. They are either dark or luminous, and their origin is unknown. Besides these casual spots, which are always small in size, there was visible during many years following 1878 a very remarkable and conspicuous large spot, strongly red in colour for several years, though it afterwards became much fainter. This spot exhibited an oval outline and was about 27,000 miles long and 8000 miles broad. For about 4 years it maintained its intense red colour and its shape almost unaltered; but after 1882, the shape remaining, the colour sensibly faded. The observations which were made on this spot during 1886 by Professor Hough at Chicago, U. S., with an 18-inch refractor, led him to the opinion that the persistence of the red spot for so many years rendered untenable the generally accepted theory that the phenomena seen on the surface of the planet are due to atmospheric causes.

Some astronomers have thought that a relationship subsists between the spots on the Sun and the spots on Jupiter. There certainly seems an apparent identity in point of time between the two classes of spots, and on the assumption that the spots on Jupiter are indicative of disturbances on the planet, Ranyard broached the idea that both classes of phenomena are dependent on some extraneous cosmical change; and are not related as cause and effect. Browning suggested many years ago that the red colour of the belts is a periodical phenomenon coinciding with the epoch of the greatest display of sunspots, but this thought does not appear to have been followed up by any one. Spots on Jupiter seem to have been first recorded by Robert Hooke in 1664. In the following year Cassini saw a spot which he found to be in motion, and by following it attentively he inferred that the planet rotated on its axis in 9h. 56m. It is a remarkable illustration of the great care bestowed by Cassini on his astronomical work that the best modern determinations of Jupiter's rotation-period differ from Cassini's estimate by only half a minute.

Bearing in mind the enormous size of Jupiter compared with the Earth, whilst its period of rotation is considerably less than half the Earth's, it will be at once seen that the velocity of matter at the planet's equator is immensely great—466 miles per minute against the Earth's 17 miles per minute. One result of this is the great intensity of the centrifugal force at the equator, and likewise the greatness of the compression of the planet's body at the poles. Hind has suggested that the great velocity which thus evidently exists may have the effect, by reason of the development of the heat which it gives rise to, of compensating the planet for the small amount of heat which owing to its distance it receives from the Sun.

On favourable occasions the brilliancy of Jupiter is very considerable; so much so that it rivals Venus and Mars. And besides this, there appears to be something special in the nature of Jupiter's surface, for not only does it seem to radiate a much larger proportion of the solar light which falls on it than do the planets generally, but some observers have expressed the opinion that it possesses inherent light of its own. Speculations, however, such as this must always be received with reserve, because of the evident difficulty of making sure of the facts on which they must be based. One thing, however, seems less open to doubt. Bearing in mind the small amount of heat which reaches Jupiter from the Sun, there is reason to infer that the clouds which certainly exist on Jupiter must owe their origin to the influence of some other heat than solar heat; in other words that Jupiter possesses sources of heat within itself.

Jupiter has satellites, 5 in number. The discovery of four of these, was one of the first fruits of the invention of the telescope, for they were found by Galileo in January, 1610. The 5th satellite is so small that it escaped notice until as recently as 1892, having been discovered on September 9 of that year by Professor Barnard, with the great Lick telescope in California. It is, however, so minute that one can count on one's fingers the telescopes capable of showing it.

The four old satellites of Jupiter shine as stars of about the 7th magnitude; in other words, they are sufficiently bright to be visible with telescopes however small: indeed several instances are on record of persons gifted with very good sight, having been able to see them with the naked eye. For the study of their physical appearance very powerful optical assistance is necessary, but their movements are so rapid, and the phenomena which result from those movements are so interesting, that these bodies may be considered to occupy the first place in the stock-in-trade of every amateur astronomer, who lays himself out for planet-gazing, with the object of profiting himself or his friends. The phenomena here alluded to are known as eclipses, transits, and occultations.

The four old satellites do not bear any names, but are numbered from the innermost outwards, and are always alluded to by their numbers as I, II, III and IV.

An eclipse of a Jovian satellite is identical in principle with an eclipse of the Moon; that is to say, just as an eclipse of the Moon happens when the Moon passes into and is lost in the Earth's shadow, so an eclipse of a Jovian satellite happens when such satellite becomes lost in the shadow cast by the planet into space. The Ist IInd and IIIrd satellites in consequence of the smallness of the inclination of their orbits, undergo eclipse once in every revolution round their primary, but the IVth is less often eclipsed, owing to the joint effect of its considerable orbital inclination, and of the distance to which it recedes from its primary.

An occultation of a Jovian satellite is akin in principle to an occultation of a star by the Moon. As the Moon moving forwards suddenly covers a star, so the planet, on occasions, suddenly covers one of its satellites. If the satellite in question is the IVth, its disappearance behind the planet and its reappearance from behind the planet will both be visible in due succession. This is often true also of the IIIrd satellite, but for reasons connected with the proximity to their primary of the Ist and IInd satellites, only their disappearance *or* reappearance (not both) can, as a rule, be observed on the same occasion. The most interesting, by far, however, of the phenomena connected with Jupiter's satellites are their transits in front of, that is across, the visible disc of the planet. Though these transits are of frequent occurrence, yet they are always interesting because of the diverse appearances which the satellites exhibit at different times, and which cannot be said to be in accordance with any recognised laws. Moreover, in observing the transit of a satellite, we may often see the black shadow cast by the satellite on the planet's disc; and this shadow will sometimes precede and sometimes follow the satellite itself. From the fact that the satellite generally appears as a bright spot on a bright background whilst the shadow is black, or blackish, an inexperienced observer is apt to look at the shadow and think he is seeing the satellite.

Jupiter revolves round the Sun in not quite 12 years at a mean distance of 483 millions of miles. Its apparent diameter varies between 50″ and 30″ according to its position with respect to the Earth. Its true diameter is about 88,000 miles. Owing to its large size and rapid rotation, as has already been mentioned, Jupiter is very much flattened at the poles. The amount of this (the polar "compression" as it is called) is about $\frac{1}{16}$.

CHAPTER X.

SATURN.

NEXT beyond Jupiter, proceeding outwards from the Sun, we reach the planet Saturn, which beyond any doubt is the most beautiful and most interesting of all the planets. Nobody who has ever had a fairly good chance of seeing it can have the least doubt that this is the case. Briefly stated the three main features which constitute its claims are:—(1) Its belts, (2) its rings, (3) its satellites.

The belts of Saturn resemble generally those of Jupiter, but they are more faint and less changeable. Their physical cause, however, may be assumed to be the same. Taking the planet as a whole, it may be said that its ordinary colour is yellowish white, the belts inclining to grayish white; though the dark belts have often been thought to exhibit a greenish hue. Lassell considered that the south pole is generally darker than the north pole and more blue in tinge.

There is one important particular in which the belts of Saturn differ from those of Jupiter. Jupiter's belts are straight, whereas Saturn's are sensibly curved. Supposing, as is probable, that Saturn's belts are parallel to the planet's equator, then we must assume that the plane of this equator makes a rather considerable angle with the ecliptic. Spots on Saturn are very rare. Whether Saturn has an atmosphere seems uncertain, or perhaps it may be said that one has not been proved to exist but may exist. The question of polar snow is also uncertain, but Sir W. Herschel

FIG. 15.—Saturn, Jan. 26, 1889 (Antoniadi).

thought he could trace changes of hue at the poles which might be due to the melting of snow.

It is usual to speak of the planet itself under

the name of the "Ball" when it is not a question of referring to the whole Saturnian system collectively. In consequence of its distance from the Sun, Saturn undergoes no equivalent to a phase; or to be more exact, no phase can be detected, though theoretically when the planet is in quadrature the disc must undergo an infinitesimally small loss of light.

Though the point has now-a-days no scientific importance, it may perhaps be desirable just to make a brief allusion to Sir W. Herschel's curious theory that Saturn was seen by him to be compressed not only at the poles but at the equator, so that it resembled a parallelogram with the corners rounded off. It is difficult to imagine what could have given rise to this strange idea, though, of course, Herschel's good faith in advancing it cannot be called in question. I refer to it because it will be found mentioned in so many books on astronomy, often under the name of the "square-shouldered" figure of Saturn. As a theory it may be regarded as quite exploded in consequence of accurate measures by Bessel, Main and others having conclusively shown that the form of the ball does not depart from that of a regular spheroid.

In referring to Saturn generally, we speak of its ring in the singular number, but, in point of fact, there are several rings—three in particular. The principal bright ring is really double, and within the innermost bright ring there is a dusky one, perfect as a ring, but not luminous as the outer rings are. By way of distinguishing one ring from another, it is usual to adopt Struve's nomenclature, whereby the outermost bright ring is called A, the inner bright ring B, and the dusky ring C.

A good engraving will convey more fully and more clearly an idea of what the Saturnian system consists of than the fullest verbal description will do. (See *Frontispiece.*)

To the earliest astronomers who possessed telescopes, Saturn proved a great puzzle, because it seemed to undergo changes of shape which were quite inexplicable on any principles then known. Galileo, when first he saw it, thought it presented an oval outline which might be due to a central planet having a smaller planet on each side of it, and accordingly he announced to his friend, Kepler, that the most distant planet was *tergeminum* or tri-form. But greater magnifying power led him to arrive at the conclusion that the planet was not a triple combination of spheres, but one body, either oblong or oval in outline. This conclusion, however, was soon found to be untenable, because the two (supposed) tributary bodies gradually decreased in size until they entirely disappeared. Galileo writing to his friend, Welser, in December 1612, thus expressed himself :—

"What is to be said concerning so strange a metamorphosis? Are the two lesser stars consumed after the manner of the solar spots? Have they vanished or suddenly fled? Has Saturn, perhaps, devoured his own children? Or were the appearances indeed illusion or fraud, with which the glasses have so long deceived me, as well as many others to whom I have shewn them? Now, perhaps, is the time come to revive the well-nigh withered hopes of those who, guided by more profound contemplations, have discovered the fallacy of the new observations, and demonstrated the utter impossibility of their existence. I do not know what to say in a case so surprising,

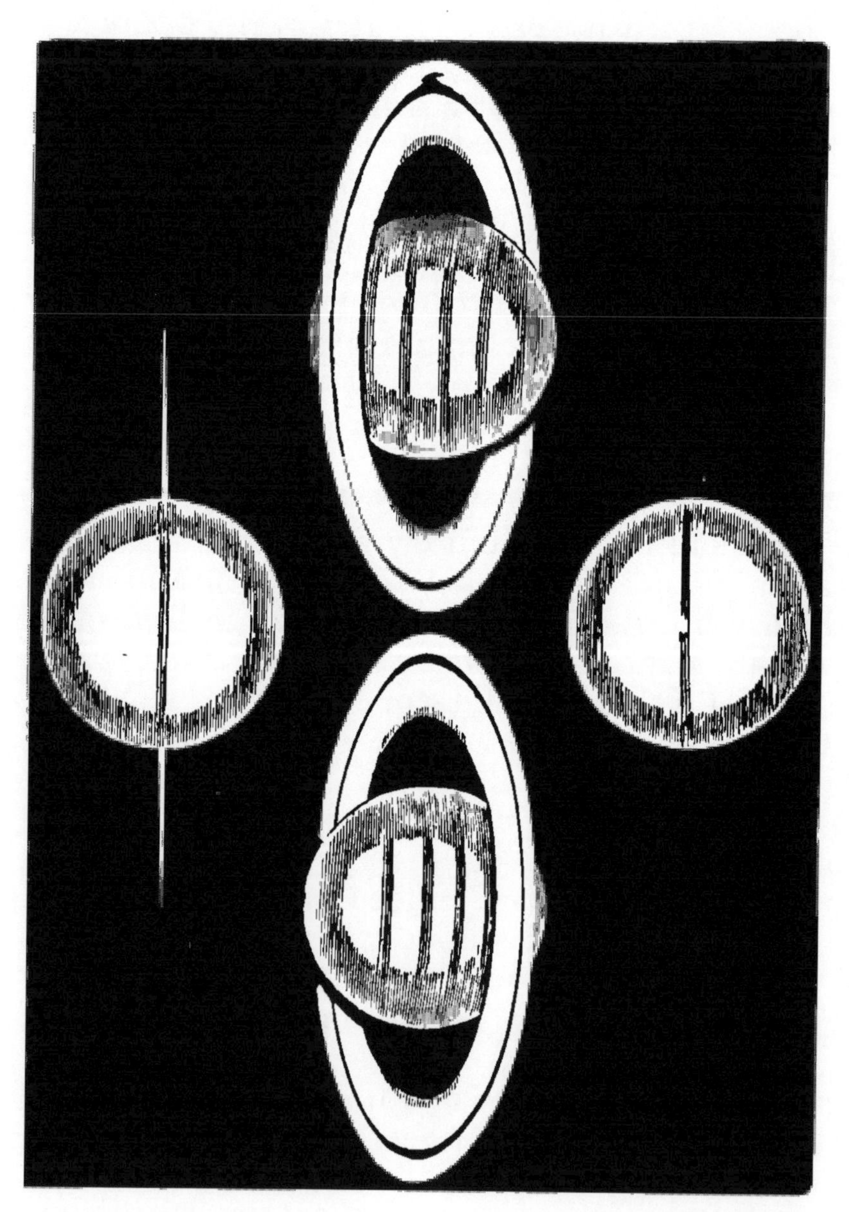

FIG. 16.—General view of the Phases of Saturn's Rings.

so unlooked for, and so novel. The shortness of the time, the unexpected nature of the event, the weakness of my understanding, and the fear of being mistaken have greatly confounded me."

Galileo seems to have become so out of heart in consequence of the difficulty of determining what these changes really meant, that he gave up altogether observing Saturn. In the course of time, but by very gradual steps, astronomers came to realise what the facts were. The next idea that was broached, was that the planet consisted of simply one central ball, and that the excrescences which Galileo had been puzzled by were merely handles as they were called, (*ansæ*) projecting like the handles, say of a soup tureen, though why they should vary in size at stated intervals remained as great a mystery as ever. It was not until about 1656 that the true explanation was arrived at by a Dutchman, named Christopher Huygens. It was the fashion in those days for scientific men to intimate to the world discoveries which they had made by resort to mysterious anagrams, which served in some degree the purpose which in the present day is served by the law regulating copyright or patent rights. Accordingly Huygens published the following singular memorandum:—

aaaaaaa cccc d eeeee g h i iiiiii llll mm nnnnnnnnn oo oo pp q rr s ttttt uuuuu.

These letters arranged in their proper order furnish the following Latin sentence:—

Annulo cingitur, tenui, plano, nusquam cohaerente, ad eclipticam inclinato; which Latin sentence becomes in the English tongue:—

"[The planet] is surrounded by a slender flat

ring inclined to the ecliptic, but which nowhere touches [the body of the planet.]"

Huygen's discovery was not a mere piece of guesswork, for he spent several years carefully observing the alterations of form which Saturn underwent, before he came to the conclusion that it was only the existence of a ring surrounding the planet which would explain the various observed changes.

It was by way of guarding himself from being robbed of the fruits of his discovery whilst he was accumulating the necessary proof of its truth, that he buried his thoughts in the logogriph or anagram just quoted. Having arrived at the conclusion which he did, he thought himself sufficiently sure of his facts to predict that in July or August 1671, the planet would again appear round, the ring becoming invisible. This surmise proved practically correct, in so far, that in May 1671, or within 2 months of the time predicted by Huygens, Cassini saw the planet as a simple ball unaccompanied by any ring.

This is a convenient place at which to offer a brief explanation of the changes of appearance as regards the ball and rings which Saturn undergoes. These changes depend jointly on Saturn's motion in its orbit round the Sun, and on the corresponding motion of the Earth in its orbit. Neither Saturn nor the Earth revolve round the Sun exactly in the ecliptic, and this want of coincidence results in the fact, that twice in the 29½ years occupied by Saturn in journeying round the Sun, the plane of its ring is seen edgeways by us on the Earth; whilst at two other periods intermediate but equi-distant the ring is seen opened out to the widest possible extent; that is, so far

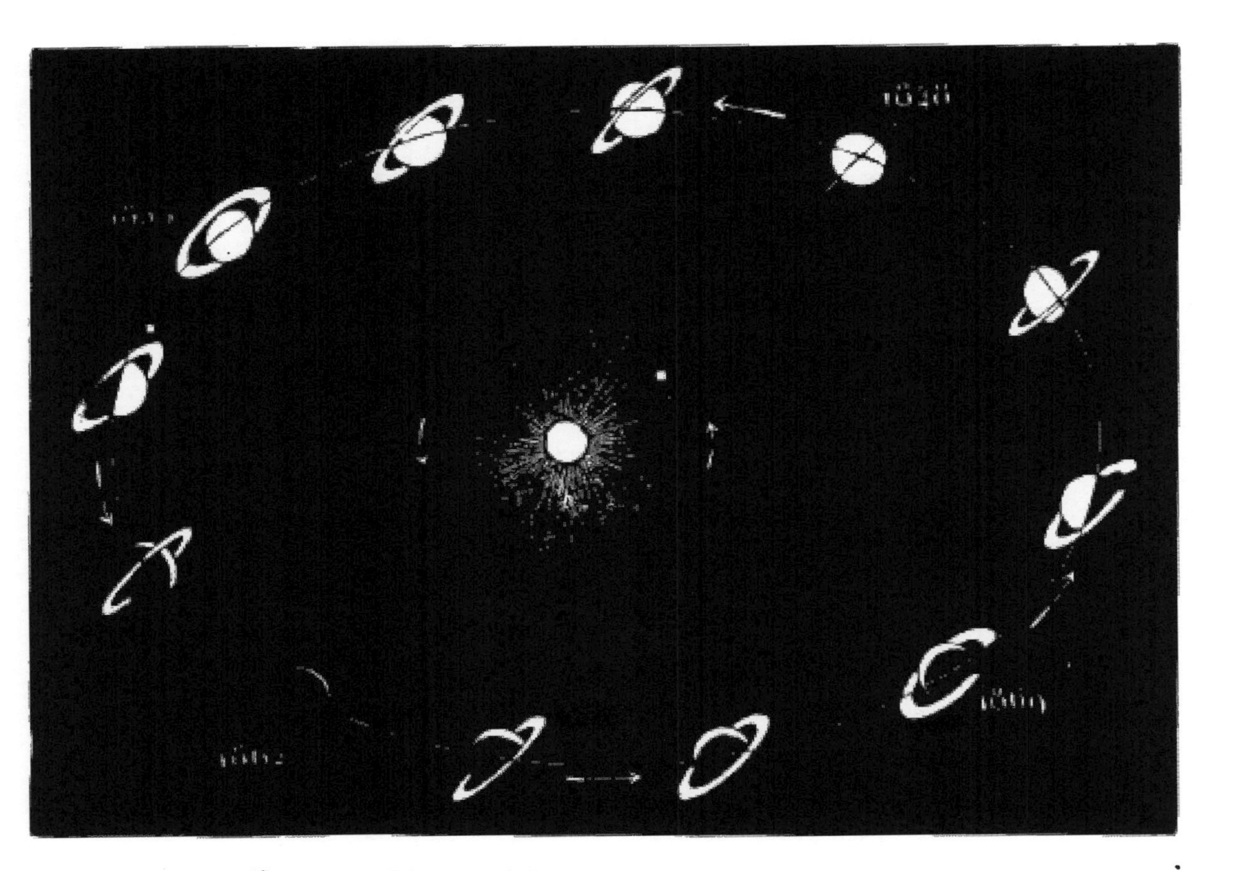

FIG. 17.—Phases of Saturn's Rings at specified dates.

as we on the Earth can by any possibility have a chance of seeing it.

The appearances presented by the rings when undergoing the transformations to which they are subject, will be readily understood by an inspection of the annexed engravings. Fig. 17, indicates the actual appearances in the years specified, and these years may be considered as carried forward and brought up to date by substituting 1877 for 1848, 1885 for 1855, 1891 for 1862, and 1898 for 1869.

Adverting to fig. 16, it will suffice to remark that the two central phases of the rings, opened wide, are to be deemed co-related, or indeed identical in a geometrical sense (so to speak) the difference being that one of them is to be deemed to show the northern side of the ring (which is now in view and will continue in view till 1907) whilst the other represents the southern side, which was in view from 1877 till 1891. The foregoing is a brief statement of the general principle involved in the changes which take place, but the motions of the two planets introduce certain technical complications into the details which would be seen by an observer using a large telescope; with these, however, the ordinary reader will not care to concern himself, and need not do so.

A great deal might be said with respect to the rings treated descriptively. I will now mention a few matters of general interest. Huygens regarded the appendage to Saturn, whose existence he established, to be a single ring, but as far back as 1675, Cassini determined that Huygen's single ring was really made up of two, one lying inside the other. Cassini in this conclusion outstepped not only all the observers of his own century, but those of the succeeding century, for Sir W. Herschel even 100 years after Cassini, was for a long time unable to satisfy himself, even with his superior telescopes, that the black streaks seen in the ring by Cassini, and regarded by him as indicative of a severance of the ring into two parts, really implied a severance. It is now, however, accepted as a fact that not only are the rings which are known as A and B absolutely distinct, but that A also is itself certainly duplex, that is, that it certainly consists of two independent rings. In addition to this many competent observers armed with powerful telescopes have obtained traces of other sub-divisions, both in A and B; and though there is some want of harmony in the details, as stated by the different observers, yet undoubtedly we must speak of Saturn's rings collectively as forming a *multiple* system.

What the rings are is a highly debatable point, but the preponderating idea is that they are not what they appear to be, namely solid masses of matter, but are swarms of independent fragments of matter. Yet "fragment" is not the best word to use, because it implies that something has been broken up to make the fragments. Rather, perhaps, we should say with Professor Young, that the rings are "composed of a swarm of separate particles, each a little independent moon pursuing its own path around the planet. The idea was suggested long ago, by J. Cassini in 1715, and by Wright in 1750, but was lost sight of until Bond revived it in connection with his discovery of the dusky ring. Professor Benjamin Pierce soon afterwards demonstrated that the rings could not be continuous solids; and Clerk Maxwell finally showed that they can be neither solid nor liquid sheets, but that all the known conditions would be answered by supposing them to consist of a flock of separate and independent bodies, moving in orbits nearly circular, and in one plane—in fact, a swarm of meteors."

The thickness of the rings seen edgeways has been variously estimated. Sir J. Herschel suggested 250 miles as an outside limit, which G. P. Bond reduced to 40 miles. It is generally considered, however, that 100 miles is probably not far from the truth. Young has pointed out that if a model of them were constructed on the scale of 1 inch to represent 10,000 miles, so that the outer ring of such a model would be nearly 17 inches in diameter, then the thickness of the ring would be represented by that of an ordinary sheet of writing paper.

Considered as a system, the rings are distinctly more luminous than the planet, and of the two bright rings, the inner one is brighter than the outer one; and the inner one is less bright at its inner edge than elsewhere. It is also to be noticed that when seen edgeways just about the time of the Saturnian equinoxes, when the Sun is shifting over from one side of the ring to the other, and the ring is dwindling down to a narrow streak, its edges (forming the *ansæ* as they are termed) do not disappear and reappear at the same time, and are not always of the same apparent extent. One ansa, indeed, is sometimes visible without the other, and most commonly it is the Eastern one that is missing. To what causes these various peculiarities are due is unknown.

Many physical peculiarities have been either noticed or suspected with reference to the bright rings. For instance, on comparing one with another, some persons have thought that their surfaces are convex, and that they do not lie in the same plane. The existence of mountains on their surface has more than once been suspected. Again, it has been fancied that they are surrounded by an extensive atmosphere. It seems hardly likely that the rings would have an atmosphere and not the ball (or *vice versa*), and, therefore, no wonder that we have no observations which countenance the idea that the ball does really possess an atmosphere. This, indeed, seems to flow from Trouvelot's observation, that the ball is less luminous at its circumference than at its centre.

The circumstances of ring C, otherwise called the "Dusky" or "Crape" ring are as curious historically, as they are mysterious physically. In 1838, Galle of Breslau, noticed what he thought to be a gradual shading off of the interior bright ring towards the ball. Though he published a statement of what he saw, the matter seems to have attracted little or no notice. In 1850, G. P. Bond in America perceived something luminous between the ring and the ball, and after repeated observations in concert with his father, came to the conclusion that the luminous appearance which he saw, was neither less nor more than an independent and imperfectly illuminated ring lying within the old rings and concentric with them. Before, however, tidings of Bond's discovery reached England, but a few days after the discovery in point of actual date, Dawes suddenly noticed one evening as Bond had done, a luminous shading within the bright rings, which he was not long in finding out to be in reality a complete ring, except so far that a portion of it was of course hidden from view behind the ball. He, and O. Struve likewise, noticed that this new Dusky Ring was occasionally to be seen divided into two or more rings. The Dusky Ring is transparent, though this fact was not ascertained until 1852, or two years after Bond's discovery of the ring.

The Dusky Ring is now recognised as a permanent feature of Saturn, but how far it used to be permanent, or how long it has been so, is a matter wrapped in doubt. Recorded observations by Picard in 1673, and by Hadley in 1723, made of course with telescopes infinitely less powerful than those of the present day, seem to suggest that both the observers named saw the Dusky Ring, without, however, being able to form a clear conception that it was a ring. It is strange that during the long period from 1723 to 1838, no one—not even Sir W. Herschel, with his various telescopes—should have obtained or at least have recorded any suspicion of its existence. There is, however, direct evidence that the Dusky Ring is wider and less faint than formerly. This was directly confirmed by Carpenter in 1863, who says he saw it "nearly as bright as the illuminated ring, so much so, that it might easily have been mistaken for a part of it." In 1883, Davidson found a marked difference in the brilliancy of the two ends (*ansæ*) of the ring.

In 1889 Barnard was fortunate enough to observe an eclipse of one of Saturn's satellites by the ring, but the eclipse, that is the concealment of the satellite, was only effected when it passed behind the bright rings; the dusky ring did not obliterate it, and hence there was obtained a conclusive proof of the transparency of the dusky ring. Barnard further concluded from his observations that there was no separating space or division between the inner bright ring and the dusky ring, as has frequently been represented in drawings. This transparency of the Dusky Ring, as a matter of fact, is therefore undoubted; yet what are we to consider to be the meaning of an observation by Wray in 1861, that whilst looking at the dusky ring edgeways the impression was conveyed to his eye that that ring was very much thicker than the bright rings?

A very interesting question which has been much discussed has reference to the stability of the rings. It is generally agreed that the constituent particles of the rings must be in motion round the primary or their equilibrium could not be maintained: almost equally certain is it, and for the like reason, that the rings cannot be solid. Of actual change in the rings as regards their dimensions, we have no satisfactory proof, though authorities differ on the point, some thinking that the rings are expanding inwards, so that ultimately they will come into contact with the ball, whilst others consider no proof whatever of such change can be obtained from any of the observations yet made in the way of measurements.

We must now proceed to consider the satellites of Saturn. These are 8 in number, 7 of which move in orbits whose planes coincide nearly with the planet's equator, whilst the remaining one is inclined about 12° thereto. One consequence of this coincidence in the planes of these satellites, which, it should be stated, are the

7 innermost, is that they are always visible to the inhabitants of both hemispheres when they are not actually undergoing eclipse in the shadow of Saturn. The satellites are of various sizes, and succeed one another in the following order, reckoning from the nearest, outwards:—Mimas, Enceladus, Tethys, Dione, Rhea, Titan, Hyperion and Iapetus. Any good 2-inch telescope will show Titan; a 3-inch will sometimes show Iapetus; a 4-inch will show Iapetus well, together with Rhea and Dione, but hardly Tethys; all the others require large telescopes. If Saturn has any inhabitants at all constituted like ourselves, which is highly improbable, they will have a chance of seeing celestial phenomena of the greatest interest. What with the rings surrounding the planet and 8 moons in constant motion, there will be an endless succession of astronomical sights for them to study. The amount of light received from the Sun cannot be much—barely $\frac{1}{100}$th of what the earth receives. The ring and satellites will therefore be useful as supplementary sources of light; yet the satellites will not furnish much, for it has been calculated that the surface of the sky occupied by all the satellites put together would to a dweller on Saturn only amount to 6 times the area of the sky covered by our Moon; whilst the intrinsic brightness of all put together would be no more than $\frac{1}{16}$th part of the light which we receive from our Moon.

The only physical fact worth noting here in connection with the satellites concerns Iapetus. Cassini two centuries ago with his indifferent telescopes thought he had ascertained that this satellite was subject to considerable variations of brilliancy. Sir W. Herschel confirmed Cassini as to this. He found that it was much less brilliant when traversing the eastern half of its orbit than at other times. Two conclusions have been drawn from this fact. One is that the satellite rotates

FIG. 18.—Saturn with the shadow of Titan on it, March 11, 1892 (Terby).

once on its axis in the same time that it performs one revolution round its primary; and that there are portions of its surface which are almost entirely incapable of reflecting the rays of the Sun. This last named supposition may perhaps be well founded, but the former needs more proof than is as yet forthcoming. Iapetus on the whole may be said to shine as a star of the 9th magnitude. To this it may be added that Titan is of the 8th magnitude, but all the others much smaller.

Saturn revolves round the Sun in a little under 29½ years at a mean distance of 886 millions of miles. Its apparent diameter varies between 15″ and 20″; its true diameter may be put at 75,000 miles. The flattening of the poles, or "polar compression" as it is called, is greater than that of any other planet, but is usually less noticeable than in the case of Jupiter, because the ring is apt to distract the eye, except when near the edgeways phase. The compression may be taken at $\frac{1}{9}$.

CHAPTER XI.

URANUS.

To the Ancients Saturn was the outermost planet of the System, nothing beyond it being known. Nor indeed was it to be assumed that any more could possibly exist, because Mercury, Venus, the Earth, Mars, Jupiter, and Saturn, with the Sun, made 7 celestial bodies of prime importance; and 7 was the number of perfection; and there was thus provided one celestial body to give a name to each of the days of the week.

But Science is not sentimental; and when men of Science come upon what looks like a discovery they do their best to bring their discovery to a successful issue, however much people's prejudices may seem to stand in the way at the moment.

On a certain evening in March, 1781, Sir William Herschel, then gradually coming into notice as a practical astronomer, was engaged in looking at different fields of stars in the constellation Gemini when he lighted on one which at once attracted his special attention. Altering his eyepiece, and substituting a higher magnifying power he found the apparent size of the mysterious object enlarged, which conclusively proved that it was not a star; for it is a well-known optical property of all stars that whatever be the size of telescope employed on them, and however high the magnifying power no definite disc of light can be obtained when in focus. Herschel's new find, therefore, was plainly not a star, and no idea having in those days come into men's minds of there being any new planets awaiting discovery, he announced as a matter of course that he had found a new comet, so soon as he ascertained that the new body was in motion. The announcement was not made to the Royal Society till April 26, more than six weeks after the date of the actual discovery, an indication, by the way, of the dilatory circulation of news a hundred years ago. The supposed comet was observed by Maskelyne, the Astronomer Royal, four days after Herschel had first seen it, and Maskelyne seems to have at once got the idea into his head that he was looking at a planet and not at a comet. As soon as possible after the discovery of a new comet the practice of astronomers is to endeavour to determine what is the shape of the orbit which it is pursuing. All attempts to carry out this in the case of Herschel's supposed new comet proved abortive, because it was found impossible to harmonise, except for a short period of time, the movements of the new body with the form of curve usually affected by most comets, namely, the parabola. It is true, as we shall see later on in speaking of comets, that a certain number of those bodies do revolve in the closed curve known as the ellipse, but it is usual to calculate the parabolic form first of all, because it is the easier to calculate; and to persevere with it until it plainly appears that the parabola will not fit in with the observed movements of the new object. This practice was carried out in the case of Herschel's new body, and it was eventually found that not only was its orbit not parabolic; that not only was its orbit not an elongated ellipse of the kind affected by comets; but that it was nearly a circle, and as the body itself showed a defined disc the conclusion was inevitable: it was in real truth a new planet. It has not taken long to write this statement, and it will take still less time for the reader to read what has been written, but the result just mentioned occupied the attention of astronomers many months in working out, step by step, in such a way as to make sure that no mistake had been made.

When it was once clearly determined that Herschel had added a new planet to the list of known planets it became an interesting matter of inquiry to find out whether it had ever been seen before; and to settle the name it should bear. A little research soon showed that the new planet had been seen and recorded as a fixed star by various observers on 20 previous occasions, beginning as far back as Dec. 13, 1690, when Flamstead registered at Greenwich as a star. These various observations, spread over a period of 91 years, and all recorded by observers of skill and eminence materially helped astronomers in their efforts to calculate accurately the shape and nature of the new planet's orbit. One observer, a Frenchman named Le Monnier, saw the planet no less than 12 times between 1750 and 1771, and if he had had (which it is known he had not) an orderly and methodical mind, the glory of this discovery would have been lost to England and obtained by France. Arago has left it on record that he was once shown one of these chance observations of Uranus, which had been recorded by Le Monnier on an old paper bag in which hair powder had been sold by a perfumer.

A long discussion took place on the question of a name for the new planet. Bode's suggestion of "Uranus" is now in universal use, but it is within the recollection of many persons living that this planet bore sometimes the name of the "Georgium Sidus" and sometimes the name of "Herschel." The former designation was proposed by Herschel himself in compliment to his sovereign and patron George III. of England; whilst a French astronomer suggested the latter name. However, neither of these appellations was acceptable to the astronomers of the Continent, who declared in favour of a mythological name, though it was a long time before they agreed to accept Bode's "Uranus." The symbol commonly used to represent the planet is formed of Herschel's initial with a little circle added below, though the Germans employ something else, "made in Germany," to quote a too familiar phrase.

The visible disc of Uranus is so small that none but telescopes of the very largest size can make anything of it. A few sentences therefore will dispose of this part of the subject. The disc is usually bluish in tinge, and most observers who look at it consider it uniformly bright, but there is satisfactory testimony to the effect that under the most favourable circumstances of instrument and atmosphere two or more belts, not unlike the belts of Jupiter, may be traced. From the position in which these belts have been seen it is inferred that the satellites of Uranus (presently to be mentioned) are unusually much inclined to the planet's equator, and revolve in a retrograde direction, contrary to what is the ordinary rule of the planets and satellites. It is assumed as the basis of these ideas, (and by analogy it is reasonable to do this) that the belts are practically parallel to the planet's equator, and at right angles to the planet's axis of rotation. To speak of the planet's axis of rotation is, in one sense, another assumption, because available observations can scarcely be said to enable us to demonstrate that the planet does rotate on its axis, yet we can have no moral doubt about it. Taylor has suggested grounds for the opinion that "there can be very little doubt that Uranus is to a very large extent self-luminous, and that we do not see it wholly by reflected light." To this Gore adds the idea that there is "strong evidence in favour of the existence of intrinsic heat in the planet."

Uranus is attended by several satellites. It was once thought that there were eight, of which six were due to Sir W. Herschel, the other two being of modern discovery. Astronomers are, however, now agreed that no more than four satellites can justly be recognised as known to exist, and they are so minute in size that only the very largest telescopes will show them; and therefore our knowledge of them is extremely limited. Sir W. Herschel's idea that he had seen six satellites appears to have resulted from his having on some occasions mistaken some very small stars for satellites. Two only of his six are thought to have been real satellites. The other two recognised satellites were found both in 1847, one by Lassell, and the other by O. Struve.

Uranus revolves round the Sun in rather more than 84 years, at a mean distance of 1781 millions of miles. Its apparent diameter, seen from the

Earth, does not vary much from 3½″, which corresponds to about 31,000 miles. It has been calculated that the light received from the Sun by Uranus would be about the amount furnished by 300 full Moons seen by us on the Earth, though another authority increases this to 1670 full Moons. From Uranus Saturn can be seen, and perhaps Jupiter, both as inferior planets, just as we see Venus and Mercury; but all the other inner planets, including Mars and the Earth, would be hopelessly lost to view, because perpetually too close to the Sun. Possibly, however, they might, on rare occasions, be seen in transit across the Sun's disc. Neptune, of course, would be visible and be the only superior planet. The Sun itself would appear to an observer on Uranus as a very bright star, with a disc of 1¾′ of arc in diameter.

CHAPTER XII.

NEPTUNE.

We now come to the best known planet of the solar system, reckoning outwards from the Sun, and though this planet itself, as an object to look at, has no particular interest for the general public, yet the history of its discovery is a matter of extreme interest. Moreover, it is very closely mixed up with the history of the planet Uranus, which has just been described. After Uranus had become fully recognised as a regular member of the solar system, a French astronomer named Alexis Bouvard set himself the task of exhaustively considering the movements of Uranus with a view of determining its orbit with the utmost possible exactness. His available materials ranged themselves in two groups:—the modern observations between 1781 and 1820, and the early observations of Flamsteed, Bradley, Mayer, and Le Monnier, extending from 1690 to 1771. Bouvard found in substance that he could frame an orbit which would fit in with each group of observations, but that he could not obtain an orbit which would reconcile both sets of observations during the 130 years over which they jointly extended. He therefore rashly came to the conclusion that the earlier observations, having been made when methods and instruments were alike relatively imperfect, were probably inaccurate or otherwise untrustworthy, and had better be rejected. This seemed for awhile to solve the difficulty, and results which he published in 1821 represented with all reasonable accuracy the then movements of the planet. A very few years, however, sufficed to reveal discordances between observation and theory, so marked and regular as to make it perfectly clear that it was not Bouvard's work which was faulty but that Uranus itself had gone astray through the operation of definite but as yet unknown causes. What these causes were could only be a matter of surmise based upon the evident fact that there was some source of disturbance which was evidently throwing Uranus out of its proper place and regular course. First one and then another astronomer gave attention and thought to the matter, and eventually the conclusion was arrived at that there existed, more remote from the Sun than Uranus, an undiscovered planet which was able to make its influence felt by deranging the movements of Uranus in its ordinary journey round the Sun every 84 years. This conclusion on the part of astronomers becoming known, a young Cambridge student, then at St. John's College, John Couch Adams by name, resolved, in July 1841, to take up the subject, though it was not until 1843 that he actually did so. The problem to be solved was to suggest the precise place in the sky at a given time of an imaginary planet massive enough to push, or pull, out of its normal place the planet Uranus, which was evidently being pushed at one time and pulled at another. It would also be part of the problem to predict the distance from the Sun of the planet thus imagined to exist. Adams worked patiently and silently at this very profound and difficult problem for 1¾ years when he found himself able to forward to Airy, who had become Astronomer Royal after being a Cambridge Professor, some provisional elements of an imaginary planet of a size, at a distance, and in a position to meet the circumstances. It is greatly to be regretted, on more grounds than one, that Airy did nothing but pigeon-hole Adams's papers. Had he done what might have been, and probably was, expected, that is, had he made them public, or better still had he made telescopic use of them, a long and unpleasant international controversy would have been avoided, and Adams would not have been robbed in part of the well-deserved fruits of his protracted labours.

We must now turn to consider something that was happening in France. In the summer of 1845, just before Adams had finished his work, and one and a half years after he commenced it a young Frenchman, who afterwards rose to great eminence, U. J. J. Le Verrier, turned his attention to the movements of Uranus with a view of ascertaining the cause of their recognised irregularity. In November 1845 he made public the conclusion that those irregularities did not exclusively depend upon Jupiter or Saturn. He followed this up in June 1846 by a second memoir to prove that an unknown exterior planet was the cause of all the trouble, and he assigned evidence as to its position very much as Adams had done 8 months previously. Airy on receiving a copy of Le Verrier's memoir seems so far, at last, to have been roused that he took the trouble to compare Le Verrier's conclusions with those of Adams so long in his possession neglected. Finding that a remarkably close accord existed between the conclusions of the two men, he came to realise that both must be of value, and he wrote a fortnight later to suggest to Professor Challis the desirability of his instituting a search for the suspected planet. Challis began within two days, but was handicapped by not having in his possession any map of the stars in the neighbourhood suggested as the *locale* of the planet. He lost no time however in making such a map, but, of course, the doing so caused an appreciable delay, and it was not until September 29, 1846, that he found an object which excited his suspicions and eventually proved to be the planet sought for. It was subsequently ascertained that the planet had been recorded as a star on August 4 and 12, and that the star of August 12 was missing from the zone observed on July 30. The discovery of the planet was therefore just missed on August 12 because the results of each evening's work were not adequately compared with what had gone before.

Meanwhile things had not been standing still in France. In August 1846, Le Verrier published a third memoir intended to develope information respecting the probable position of the planet in the heavens. In September 23 a summary of this third memoir was received by Encke at Berlin, accompanied by the request that he would co-operate instrumentally in the search for it. Encke at once directed two of his assistants named D'Arrest and Galle to do this, and they were fortunately well circumstanced for the task. Unlike Challis, who, as we have seen, could do nothing until he had made a map for himself, the Berlin observers had one ready to hand, which by good chance had just been published by the Berlin Academy for the part of the heavens which both Adams and Le Verrier assigned as the probable locality in which the anxiously desired planet would be found. Galle called out the visible stars one by one whilst D'Arrest checked them by the map, and suddenly he came upon an unmarked object which at the moment looked like an 8th magnitude star. The following night showed that the suspicious object was in motion, and it was soon ascertained to be the trans-Uranian planet which was being searched for. The discovery when announced excited the liveliest interest all over the world. It did more; it created a bitter feeling of resentment on the part of French astronomers that the laurels claimed by them should have been also claimed in an equal share by a young and unknown Englishman, and accordingly the old cry of "*perfide Albion*" arose on all sides. I have been particular in stating the various dates which belong to this narrative, in order to make as clear as possible the facts of the case. This is even now necessary, because though the astronomers of England and Germany are willing to give Adams and Le Verrier each their fair share of this great discovery, the same impartial spirit is not to be found in France, for nothing is more common, even in the present day, in looking at French books of astronomy, than to find Adams's name either glossed over or absolutely suppressed altogether when the planet Neptune is under discussion.

How remarkable a discovery this was, will perhaps be realized, when it is stated that Adams was only $2\frac{1}{2}°$ out in assigning the position of the new planet, whilst Le Verrier was even nearer, being barely $1°$ out.

We know practically nothing respecting the physical appearance of Neptune, owing to its immense distance from us, and for the like reason the Neptunian astronomers, if there are any, will know absolutely nothing about the Earth; indeed, their knowledge of the Solar System will be restricted to Uranus, Saturn, and the Sun. Even the Sun will only have an apparent diameter of about $1'$ of arc, and, therefore, will only seem to be a very bright star, yielding light equal in amount, according to Zollner, to about 700 full moons. There is one satellite belonging to Neptune, and as this has been calculated to exhibit a disc $10°$ in diameter, a certain amount of light will no doubt be afforded by it especially if, as is not unlikely, Neptune is itself possessed of some inherent luminosity independently of the Sun.

The fact that Neptune seems destitute of visible spots or belts, results in our being ignorant of the period of its axial rotation, though it should be stated that in 1883, Maxwell Hall in Jamaica, observed periodical fluctuations in its light, which he thought implied that the planet rotated on its axis in rather less than 8 hours. Several observers thought 20 or 30 years ago, that they had noticed indications of Neptune being surrounded by a ring like Saturn's ring, but the evidence as to this is very inconclusive. It is quite certain that none but the very largest telescopes in the world would show any such appendage, and this planet seems to have been neglected of late years, by the possessors of such telescopes. Moreover, if a ring existed it would only open out to its full extent once in every 82 years, being the half of the period of the planet's revolution round the Sun (just as Saturn's ring only opens out to the fullest extent every 14½ years), so that, obviously, supposing suspicions of a ring dating back 30 or 40 years were well founded, it might well be that another 30 or 40 years might need to elapse, before astronomers would be in a position to see their suspicions revive.

Neptune revolves round the Sun in 164½ years, at a mean distance of 2791 millions of miles. Its apparent diameter scarcely varies from 2¾″. Its true diameter is about 37,000 miles. No compression of the Poles is perceptible. Its one satellite revolves round Neptune in 5¾ days, and in a retrograde direction, at a mean distance of 223,000 miles, and shines as a star of the 14th magnitude. This is a peculiarity which it only shares with the satellites of Uranus, so far as it regards the planetary members of the Solar System, though there are many retrograde Comets.

The question has often been mooted, whether there exists, and belonging to the Solar System, a planet farther off than Neptune. There does seem some evidence of this, as we shall better understand, when we come to the subject of long-period Comets, though it cannot be said that much progress has yet been made in arriving at a solution of the problem.

Unless there does exist a trans-Neptunian planet, a Neptunian astronomer will know very little about planets, for Uranus and Saturn will alone be visible to him. Both will of course be what we call "inferior planets," and under the best of circumstances will cut a poor figure in the Neptunian sky.

CHAPTER XIII.

COMETS.

I SUPPOSE that it is the experience of all those who happen to be in any sense, however humble, specialists in a certain branch of science, that from time to time, they are beset with questions on the part of their friends respecting those particular matters which it is known that they have specially studied. There is no fault to be found with this thirst for information, always supposing that it is kept within due bounds; but my motive for alluding to it here, is to see whether any well-marked conclusion can be drawn from it, within my own knowledge as regards astronomical facts or events. Now in the case of the science of astronomy (for which in this connection I, for the moment, will venture to speak), there is certainly no one department which so unfailingly, at all times and in all places, seems to evoke such popular sympathy and interest as the department which deals with Comets.

Sun-spots may come and go; bright planets may shine more brightly; the Sun or Moon may be obscured by eclipses; temporary stars may burst forth,—all these things are within the ken of the general public by means of newspapers or almanacs, but it is a comet which evokes more questionings and conversations than all the other matters just referred to put together. When a new and bright comet appears, or even when any comet not very bright gets talked about, the old question is still fresh and verdant—"Is there any danger to the Earth to be apprehended from collision with a Comet?" followed by "What is a Comet?" "What is it made of?" "Has it ever appeared before?" "Will it come back again?" and so on. Questions in this strain have more often than I can tell of been put to me. They seem the stock questions of all who will condescend to replace for five minutes in the day the newest novel or the pending parliamentary election.

It may be taken as a fact (though in no proper sense a rule) that a bright and conspicuous comet comes about once in 10 years, and a very remarkable comet every 30 years. Thus we have had during the present century bright comets in 1811, 1825, 1835, 1843, 1858, 1861, 1874 and 1882, whereof those of 1811, 1843, and 1858 were specially celebrated. Tested then by either standard of words "bright and conspicuous," or "specially celebrated," it may be affirmed that a good comet is now due, so let us prepare for it by getting up the subject in advance.

I will not attempt to answer in regular order or in any set form the questions which I have just mentioned as being stock questions, but they will be answered in substance as we go along. There is one matter in connection with comets which has deeply impressed itself upon the public mind, and that is the presence or absence of a "tail." It is not too much to say that the generality of people regard the tail of a comet as *the* comet; and that though an object may be a true comet from an astronomer's point of view, yet if it has no tail its claims go for nought with the mass of mankind. We have here probably a remnant of ancient thought, especially of that line of thought which in bygone times associated Comets universally with the idea that they were especially sent to be portents of national disasters of one kind or another. This is brought out by numberless ancient authors, and by none more forcibly than Shakespeare. Hence we have such passages as the following in *Julius Cæsar* (Act ii., sc. 2):—

> "When beggars die there are no comets seen,
> The Heavens themselves blaze forth the death of princes."

In *Henry VI.* (Part I., act i., sc. 1) we find the well-known passage:—

> "Comets importing change of times and states
> Brandish your crystal tresses in the sky,
> And with them scourge the bad revolting stars
> That have consented unto Henry's death."

There are in point of fact two distinct ideas evolved here: (1) that comets are prophetic of evil, and (2) stars potential for evil.

There is another passage in *Henry VI.* (Part I., act iii., sc. 3) even more pronounced:—

> "Now shine it like a Comet of revenge,
> A prophet to the fall of all our foes."

Again; in *Hamlet* (Act i., sc. 1) we find:—

> "As stars with trains of fire, and dews of blood,
> Disasters in the Sun."

Once more; in the *Taming of the Shrew* (Act iii., sc. 2) we have the more general, but still emphatic enough, idea expressed by the simple words of reference to—

> "Some Comet or unusual prodigy."

Shakespeare may be said to have lived at the epoch when astrology was in high favour, and it may be that he only gave utterance to the current opinion prevalent among all classes in those still somewhat "Dark Ages" (so called). This, however, can hardly be said of the author of my next quotation—John Milton (*Paradise Lost*, bk. II.):—

> "Satan stood
> Unterrified, and like a Comet burned,
> That fires the length of Ophiuchus huge
> In th' Arctic sky, and from its horrid hair
> Shakes pestilence and war."

Jumping over a century we find the ancient theory still in vogue, or Thomson (*Seasons*, Summer) would never have written:—

> "Amid the radiant orbs
> That more than deck, that animate the sky,
> The life-infusing suns of other worlds;
> Lo! from the dread immensity of space,
> Returning with accelerated course,
> The rushing comet to the sun descends;
> And, as he sinks below the shading earth,
> With awful train projected o'er the heavens,
> The guilty nations tremble."

Even Napoleon I. had servile flatterers who, as late as 1808, tried to extract astrological influence out of a comet by way of bolstering up "Old Bony." But enough of poetry and fiction, let us hasten back to prosaic fact.

Comets as objects to look at may be classed under three forms, though the same comet may undergo such changes as will at different epochs in its career cause it to put on each variety of form in succession. Thus the comet of 1825 seen during that year as a brilliant naked-eye object, after being lost in the sun's rays, was again found on April 2, 1826 by Pons. Lamentable were his cries at the miserable plight it was in. He described it as totally destroyed: without tail, beard, coma or nucleus, a mere spectre. The simplest form of comet is a mere nebulous mass, almost always circular, or perhaps a little oval, in outline. It may maintain this appearance throughout its visibility; or, growing brighter may become a comet of the second class, with a central condensation, which developing becomes a "nucleus" or head. It may retain this feature for the rest of its career, or may pass into the third class and throw out a "coma" or beard, which will perhaps develop into a tail or tails. Doing this it will not unfrequently grow bright enough and large enough to become visible to the naked eye. In exceptional cases the nucleus will become as bright as a 2nd or even 1st mag-

FIG. 19.—T cope with a nucleus

nitude star, and the tail may acquire a length of several or many degrees. In the last named case of all the comet becomes, *par excellence* according to the popular sentiment, "a comet." It will now be readily inferred that the astronomer in his observatory has to do with many comets which the public at large never hear of, or if they do hear of, treat with contempt, because they are destitute of tails.

The tails of comets exhibit very great varieties not only of size but of form; some are long and

FIG. 20.—Wells's Comet of 1882, seen in full daylight near the Sun on Sept. 18.

slender; some are long and much spread out towards their ends, like quill pens, for instance; some are short and stumpy, mere tufts or excrescences rather than tails. Not unfrequently a tail seems to consist of two parallel rays with no cometary matter, or it may be only a very slight amount of cometary matter traceable in the in-

FIG. 21.—Quenisset's Comet, July 9, 1893 (Quenisset).

terspace; some have one main tail consisting of a pair of rays such as just described, together with one or more subsidiary or off-shoot tails. The comet of 1825 had five tails and the comet of 1744 had six tails. It might be inferred from all this that the tails of comets are so exceedingly irregular, uncertain and casual as to be amenable to no laws. This was long considered to be the case; but a Russian observer named Bredichin, as the result of much study and research, has arrived at the conclusion that all comet tails may be brought under one or other of three types; and that each type is indicative of certain distinct differences of origin and condition which he considers himself able to define. The first type comprises tails which are long and straight; "they are formed" (to quote Young's statement of Bredichin's views) "of matter upon which the Sun's repulsive action is from twelve to fifteen times as great as the gravitational attraction, so that the particles leave the comet with a relative velocity of at least four or five miles a second; and this velocity is continually increased as they recede, until at last it becomes enormous, the particles travelling several millions of miles in a day. The straight rays which are seen in the figure of the tail of Donati's Comet, tangential to the tail, are streamers of this first type; as also was the enormous tail of the comet of 1861. The second type is the curved plume-like train, like the principal tail of Donati's Comet. In this type the repulsive force varies from 2.2 times gravity (for the particles on the convex edge of the tail) to half that amount for those which form the inner edge. This is by far the most common type of cometary train. A few comets show tails of the third type—short, stubby, brushes violently curved, and due to matter of which the repulsive force is only a fraction of gravity—from $\frac{1}{10}$ to $\frac{1}{2}$."

Bredichin wishes it to be inferred that the tails of the 1st type are probably composed of hydrogen; those of the 2nd type of some hydrocarbon gas; and those of the 3rd of the vapour of iron, probably with some admixture of sodium and other substances. Bredichin, as a reason for these conclusions, supposes that the force which generates the tails of comets is a repulsive force, with a surface action the same for equal surfaces of any kind of matter; the effective accelerating force therefore measured by the velocity which it would produce would depend upon the ratio of surface to mass in the particles acted upon, and so, in his view, should be inversely proportional to their molecular weights. Now it happens that the molecular weights of hydrogen, of hydro-carbon gases, and of the vapour of iron bear to each other just about the required proportion.

I am here stating the views and opinions of others without definitely professing to be satisfied with them, but as they have met with some acceptance, it is proper to chronicle them, though we know nothing of the nature of the repulsive force here talked about. It might be electric, it might be anything. The spectroscope certainly lends some countenance to Bredichin's views, but we need far more knowledge and study of comets before we shall be justly entitled to dogmatise on the subject.

This has been rather a digression. I go back now to prosaic matters of fact, of which a vast and interesting array present themselves for consideration in connection with comets. Let us consider a little in detail what they are, to look

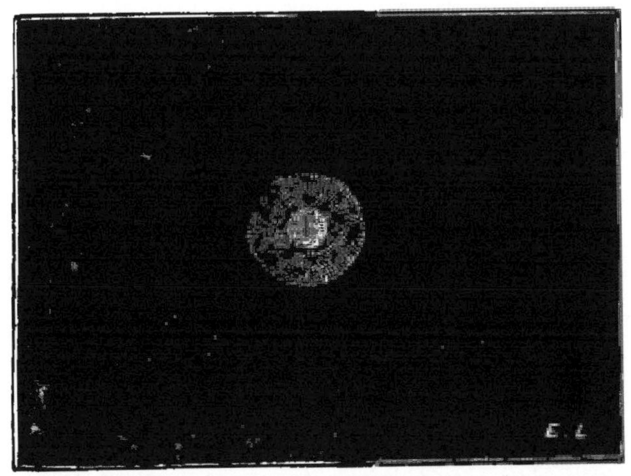

FIG. 22.—Holmes's Comet, Nov. 9, 1892 (Denning).

at. We have seen that a well-developed comet of the normal type usually comprises a nucleus, a head or coma, and a tail. Comets which have

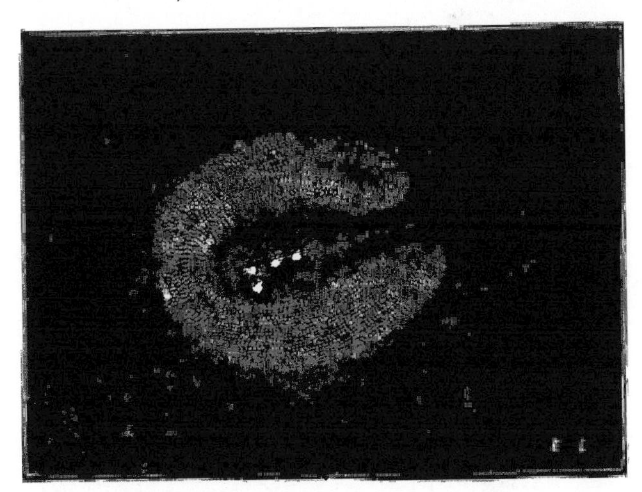

FIG. 23.—Holmes's Comet, Nov. 16, 1892 (Denning).

no tails generally exhibit heads of very simple structure; and if there is a nucleus, the nucleus is little else than a stellar point of light. But in the case of the larger comets, which are almost or quite visible to the naked eye, the head often exhibits a very complex structure, which in not a few cases seems to convey very definite indications of the operations going on at the time. Figs. 22 and 23 may be taken as samples of a complex cometary head, though no two comets resemble one another exactly in details. Fig. 24 forcibly conveys the idea that we are looking at a process of development analogous to an uprush of water from a fountain, or perhaps I might better say, from a burst waterpipe. There is a distinct idea of a jet. This self-same idea, in another form, presents itself in the case of those comets which exhibit what astronomers are in the habit of calling "luminous envelopes." The jet in this case is not strictly a jet because it is not a continuous outflow, or overflow, of matter; the idea rather suggests itself of an intermittent overflow resulting in accumulated layers, or strata, of matter becoming visible. But with this we come to a standstill; we cannot tell where the matter comes from and still less where it goes to; we can only record what our eyes, assisted by telescopes, tell us. There can, however, I think, be no doubt that the matter of a comet becomes displayed to our senses as the result of a process of expulsion, or repulsion, from the nucleus; and then, having become launched into space, it comes under the

FIG. 24.—Comet III. of 1862, on Aug 22, showing jet of luminous matter (Challis).

influence, also repulsive, of the Sun. All these things are visible facts. As to causes, we suggest little, because we know so little. Anyone who has seen a comet and has watched the displays of jets and luminous envelopes, such as I have endeavoured to set forth, will realise at once how impossible it is to describe these things in words. They must be seen either in actual being or in picture. Some further allusions to this branch of the subject may perhaps be more advantageously made after we have considered the movements and orbits of comets.

There is often a slight general resemblance between a planet and a comet, as regards the path which each class of body pursues. Probably the least reflective person likely to be following me here understands the bare fact, that all the planets revolve round the Sun, and are held to defined orbits by the Sun's influence, or attraction, as it is called. Perhaps, it is not equally realised, that in a somewhat similar, but not quite the same way, comets are influenced and controlled by the Sun.

Comets must be considered as regards their motions to be divisible into two classes:—(1) Those which belong to the Solar System; and (2) those which do not. Each of these two classes must again be sub-divided, if we would really obtain a just conception of how things stand.

By the Comets which belong to the Sun, I mean those which revolve round the Sun in closed orbits;* and are, or may be, seen again and again at recurring intervals. There are 2

* The circle and the ellipse are what are called "closed" curves.

or 3 dozen comets which present themselves to our gaze at stated intervals, varying from about 3 to 70 years. There are again other comets which without any doubt (mathematically) are revolving round the Sun in closed orbits, but in orbits so large and with periods of revolution so long (often many centuries), that though they will return again to the sight of the inhabitants of the earth some day, yet no second return having been actually recorded, the astronomer's prediction that they will return, remains at present a prediction based on mathematics but nothing more.

There is another class of Comet of which we see examples from time to time, and having seen them once shall never see again. This is because these Comets move in orbits which are not closed, and which are known as parabolic or hyperbolic orbits respectively, because derived from those sections of a cone which are called the Parabola and the Hyperbola. It must be understood that what I am now referring to is purely a matter of orbit, and that no relationship subsists between the size and physical features of a Comet and the path it pursues in space. The only sort of reservation, perhaps, to be made to this statement is, that the comets celebrated for their size and brilliancy, are often found to be revolving in elliptic orbits of great eccentricity, which means that their periods may amount to many centuries.

It may be well to say something now as to what is the ordinary career of a comet, so far as visibility to us, the inhabitants of the Earth, is concerned. Though this might be illustrated by reference to the history of many comets, perhaps there is no one more suitable for the purpose than Donati's Comet of 1858. In former times, when telescopes were few or non-existent, brilliant comets often appeared very suddenly, just as a carriage or a man does, as you turn the corner of a street. Such things even happen still: for instance, the great comet of 1861 burst upon us all at once at a day's notice. Usually, however, now in consequence of the large size of the telescopes in use, and the great number of observers who are incessantly on the watch, comets are discovered when they are very small, because remote both from the Earth and Sun, and many weeks, or even months, it may be, before they shine forth in their ultimate splendour. Now, let us see how these statements are supported by the history of Donati's comet in 1858. On June 2 in that year, it was first seen by Donati at Florence, as a faint nebulosity, slowly journeying northwards. June passed away, and July, and August, the comet all the while remaining invisible to the naked eye; that is to say, it first became perceptible to the naked eye on August 29, having put forth a faint tail about August 20. After the beginning of September its brilliancy rapidly increased. On September 17, the head equalled in brightness a 2nd magnitude star, the tail being 4° long. Passing its point of nearest approach to the Sun on September 29, it came nearest to the Earth on October 10; though, perhaps, its appearance a few days previously, namely on October 5, is the thing best remembered by those who saw it, because it was on that night that the comet passed over the 1st magnitude star Arcturus. For several days about this time, the comet was an object of striking beauty in the Western

Heavens, during the hours immediately after sun-set. After October 10, it rapidly passed away to the Southern hemisphere, diminishing in brightness, as it did so, because receding from the Earth and the Sun. It continued its career through the winter; became invisible to the naked eye; and finally invisible altogether in March 1859. It remained in view, therefore, for more than nine months, not to return again till about the year 3158 A. D., for its period of revolution was found to be about 2000 years.

I have been particular in sketching somewhat fully the history of this comet so far as we are concerned, because, as I have already said, it is typical of the visible career of many comets. Halley's comet in 1835 and 1836, went through a somewhat similar series of changes. This comet—a well-known periodical one of great historic interest and brilliancy—may be commended to the younger members of the rising generation, because it is due to return again to these parts of space a few years hence, or in 1910.

What is a comet made of? Men of Science equally with the general public would like to be able to answer this question, but they cannot do so with satisfactory certainty. A great many years ago Sir John Herschel wrote thus:—"It seems impossible to avoid the following conclusion, that the matter of the nucleus of a comet is powerfully excited and dilated into a vaporous state by the action of the Sun's rays escaping in streams and jets at those points of its surface which oppose the least resistance, and in all probability throwing that surface or the nucleus itself into irregular motions by its reaction in the

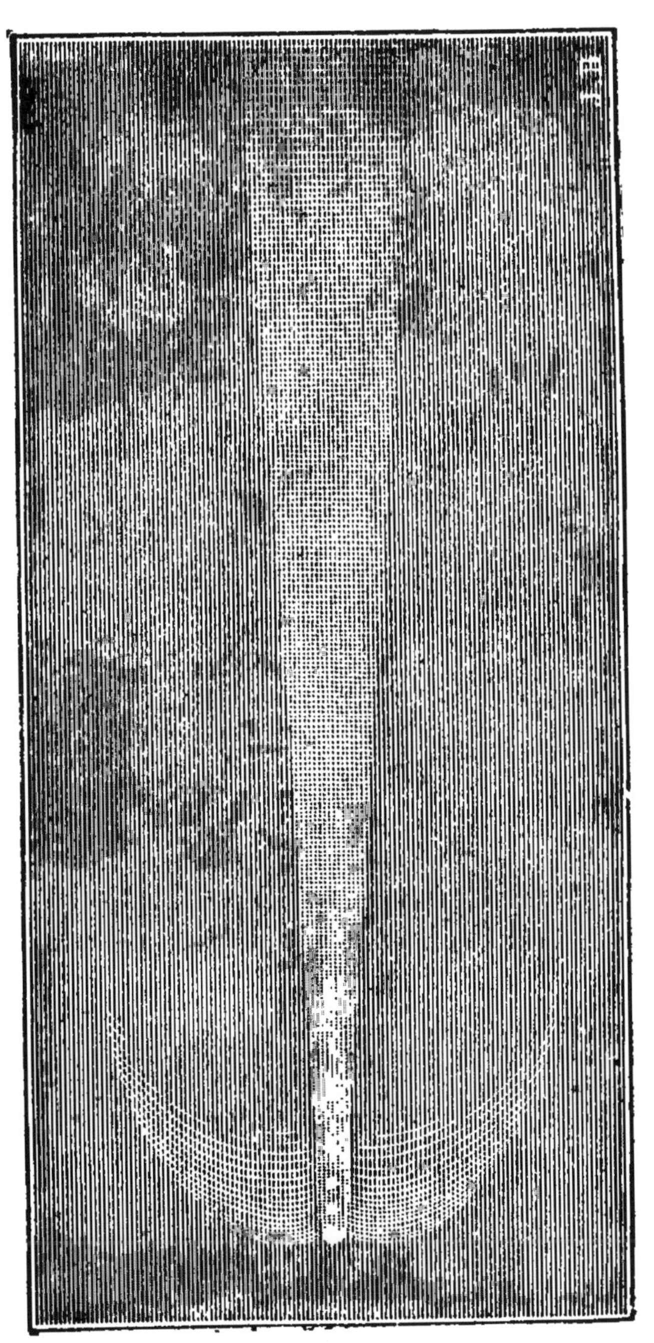

FIG. 25.—Sawerthal's Comet, June 4, 1888 (Charlois).

act of so escaping, and thus altering its direction." This passage was written of course before the spectroscope had been brought to bear on the observations of comets, but so far as Sir John Herschel's remark implies the presence of vapour, that is gas, in a comet, the surmise has been amply borne out by later discoveries. The fact that as a comet approaches the Sun some forces, no doubt of solar origin, come into operation to vaporise and therefore expand the matter composing the comet is sufficiently shown by the great developement which takes place as we have seen in the tails of comets, but in regard to the heads of comets we are face to face with a strange enigma. Though the tails expand the heads contract as the comet approaches its position of greatest proximity to the Sun. Having passed this point the head expands again. This curious circumstance, first pointed out by Kepler in 1618, has often been noticed since, and noticed indeed not as the result of mere eye impressions, but after careful micrometrical measurement with suitable instruments. I think the confession must be made that we are hopelessly ignorant of the nature of comet's except that gases are largely concerned in their constitution.

It seems impossible to doubt that some tails of comets are hollow cylinders or hollow cones. Such a theory would account for the fact, so often noticed, that single tails are usually much brighter at their two edges than at the centre. This is the natural effect of looking transversely at any translucent cylinder of measureable thickness.

It was long a moot point whether comets are self-luminous, or whether they shine by reflected light; but it is now generally admitted that whilst a part of the light of a comet may be derived by reflection from the Sun yet as a rule they must be regarded as shining by their own intrinsic light.

It should be stated here by way of caution that the observations on this subject are not so consistent as one could wish, and it seems necessary to assume that all comets are not constituted alike, and that therefore what is true of one does not necessarily apply to another.

To those who possess telescopes (not necessarily large ones) opportunities for the study of comets have much multiplied during the last few years, for we are now acquainted with a group of small comets which are constantly coming into view at short intervals of time. The comets have now become so numerous that seldom a year passes without one or more of them coming into view. Whilst that known as Encke's revolves round the Sun in 3¼ years, Tuttle's doing the same in 13½ years, there are four others whose periods average about 5½ years, 5 which average 6½ years, together with one of 7½ years and one of 8 years. It is thus evident that there is a constant succession of these objects available for study, and that very few months can ever elapse that some one or more of them are not on view. They bear the names of the astronomers who either discovered them originally, or who, by studying their orbits, discovered their periodicity. The names run as follows, beginning with the shortest in period and ending with the longest:—

Encke's.
Temple's Second (1873, II.)
Winnecke's.
Brorsen's.
Temple's First (1867, II.)

Swift's (1880, v.)	Wolf's (1884, III.)
Barnard's (1884, II.)	Faye's.
D'Arrest's.	Denning's.
Finlay's.	Tuttle's.

I cannot stay to dwell upon either the history or description of these comets separately, but must content myself by saying generally that whilst as a rule they are not visible to the naked eye, yet several of them may occasionally become so visible when they return to perihelion under circumstances which bring them more near than usual to the earth.

Several other comets are on record which it was supposed at one time would certainly have been entitled to a place in the above list, but three of them in particular have, under very mysterious circumstances, entirely disappeared from the Heavens.

Chief amongst the mysterious comets must be ranked that which goes by the name of Biela. This comet, first seen in 1772, was afterwards found to have a period of about 6¾ years, and on numerous occasions it reappeared at intervals of that length down to 1845, when the mysterious part of its career seems to have commenced. In December of that year this comet threw off a fragment of nearly the same shape as itself, and the two portions travelled together side by side for four months, the distance between the fragments slowly increasing. At the end of the four months in question the comet passed out of sight owing to the distance from the earth to which it had attained. The comet returned again to perihelion in 1852, remaining visible for three weeks. The two portions of the comet noticed in 1846 retained their individuality in 1852, but the distance between them had increased to about eight times the greatest distance noticed in 1846. As a comet Biela's Comet has never been seen since 1852, and it must now be regarded as having permanently disappeared. But what seems to have happened is this, that Biela's Comet has become

FIG. 26.—Biela's Comet, February 19, 1846.

broken up into a mass of meteors. On November 27, 1872, and again in November 1885, when the earth in travelling along its own orbit reached a certain point where its orbit intersected the former orbit of Biela's Comet the Earth encountered, instead of the comet which ought to have been there, a wonderful mass of meteors; and it is now generally accepted that these meteors, which apparently are keeping more or less together as a fairly compact swarm, are nought else than the disintegrated materials of what once was Biela's Comet.

It is extremely probable that as time goes on we shall be able to say that an intimate connection subsists between particular comets which have been and particular meteoric swarms. We already possess proof that other comets which once came within our view were at that time revolving round the Sun in orbits so comparatively small that they should have reappeared at intervals of half-a-dozen or so years, yet they have not reappeared. The question therefore suggests itself, Have they been subject to some great internal disaster which has led to their disintegration? It may be said without doubt that this is in the highest degree probable; but short of this, that is short of total disintegration into small fragments, we have several cases on record of what I may, for the moment, call ordinary comets breaking up into two or three fragments. For a long while astronomers were naturally loath to believe that this was possible, and therefore they discredited the statements to that effect which had been made. Though it would occupy too much space to give the particulars of these comets in full it may yet be worth while just to mention the names of some of them, presumed to be of short period, which seemed nevertheless to have eluded our grasp. I would here specially mention Liais's Comet of 1860 and the second comet of 1881 as seemingly having undergone some sort of disruption akin to what happened in the case of Biela's Comet.

There is another group of periodical comets

to be mentioned. These are six in number and seem to have periods of 70 years or a little more. Of these three have not yet given us the chance of seeing them again; two have paid us a second visit, and therefore their periods are not open to doubt; whilst the most famous of this group, "Halley's," has been recorded to have shown itself to the Earth no less than 25 times, beginning with the year 11 B. C. It was Halley's comet which shone over Europe in April 1066, and was considered the forerunner of the conquest of England by William of Normandy. It figures in the famous Bayeux tapestry as a hairy star of strange shape.

It would seem that there exists in some inscrutable manner a connection between each of the three great exterior planets and certain groups of comets. In the case of Jupiter the association is so very pronounced as long ago to have attracted notice; but the French astronomer, Flammarion, has brought forward some suggestions that Saturn has one comet (and perhaps two) with which it is associated; Uranus, two (and perhaps three); and Neptune, six; whilst farther off than Neptune the fact that there are two comets, supposed periodical, without a known planet to run with them has inspired Flammarion to look with a friendly eye on the idea (often mooted) that outside of Neptune there exists another undiscovered planet revolving round the sun in a period of about 300 years.

The Jupiter group of comets deserves a few additional words. There are certainly nine, and perhaps twelve comets revolving round the Sun in orbits of such dimensions that they either reach up to or slightly overreach the orbit of

Jupiter. The effect of this condition of things is that on occasions Jupiter and each of the comets may come into such proximity that the superior mass of Jupiter may exercise a very seriously disturbing influence over a flimsy and ethereal body like a comet. There is reason to suppose that some of the comets now belonging to the Jupiter group have not done so for any great length of time, but having been wandering about, either in elliptic orbits of great extent, or even in parabolic orbits, have accidentally come within reach of Jupiter, and so have been, as it were, captured by him. Hence the origin of the term, the "capture theory," as applied to these family groups of comets which I have just stated to exist, each presided over, as it were, by a great planet. It may be that at some future time this theory will help us to a clue to the fact that besides the comets of Lexell of 1770, Blainpain of 1819, and Di Vico of 1844, short period comets unaccountably missing, there are several others presumed to have been revolving in short period orbits when discovered, and as to which it is very strange that they should not have been seen before their only recorded visit to us, and equally strange that they should never have been seen since.

Is there any reason to fear the results of a collision between a comet and the Earth? None whatever. However vague may be, and in a certain sense must be, our answer to the question, "What is a comet?" certain is it that every comet is a very imponderable body—a sort of airy nothing, a mass of gas or vapour.* At the

* It is not a little singular that the Chinese in bygone centuries have often alluded to comets under the name of

same time it always has been and perhaps still is difficult to persuade the public that whatever might be the effect on a comet if it were to strike the Earth, the effect on the Earth, were it to be struck by a comet, would be *nil*. This is not altogether a matter of speculation, for according to a calculation by Hind, on June 30, 1861, the Earth passed into and through the tail of the great comet of that year at about two-thirds of its distance from the nucleus. Assuredly there was no dynamical result; but it seems, however, not unlikely that there was an optical result; at any rate, traces of something of this sort were noted. Hind himself, in Middlesex, observed a peculiar phosphorescence or illumination of the sky which he attributed at the time to an auroral glare. Lowe, in Nottinghamshire confirmed Hind's statement of the appearance of the heavens on the same day. The sky had a yellow auroral glare-like look, and the Sun, though shining, gave but feeble light. The comet was plainly visible at 7.45 p. m. (during sunshine), and had a much more hazy appearance than on any subsequent evening. Lowe adds that his Vicar had the pulpit candles lighted in the Parish Church at 7 o'clock (it was a Sunday), though only five days had elapsed since Midsummer day, which itself proves that some sensation of darkness was felt even while the Sun was shining.

So far as I remember there has been no such thing as a comet panic during the present generation, at any rate in civilised countries, but it is on record that there was a very considerable

---

vapours; *e. g.*, the comet of 1618 is recorded as having been "a white vapour 20 cubits long."

panic in 1832 in connection with the return of Biela's Comet in the winter of that year. Olbers as the result of a careful study in advance of the comet's movements found that the comet's centre would pass only 20,000 miles within the Earth's orbit, and that as the nebulosity of the comet had in 1805 been more than 20,000 miles in diameter, it was certain, unless its dimensions had diminished in the 27 years, that some of the comet's matter would overlap the Earth's orbit; in other words would envelop the Earth itself, if the Earth happened to be there. This conclusion when it became public was quite enough to create a panic and make people talk about the forthcoming destruction of our globe. It was nothing to the point (in the public mind) that astronomers were able to predict that the Earth would not reach the place where the comet would cross the Earth's orbit until four weeks after the comet had come and gone. However, we now know that nothing happened, and I am justified in adding that even if there had been contact, Earth meeting comet face to face, nothing (serious) would have occurred so far as the Earth was concerned.

This seems a convenient place for referring to a matter which when it was first broached excited a great deal of interest, but about which one does not hear much now-a-days. The period of the small comet known as Encke's (which, revolving as it does round the Sun in a little more than three years, has the shortest period of any of the periodical comets) was found many years ago to be diminishing at each successive return. That is to say, it always attained its nearest distance from the Sun at each apparition 2½ hours sooner than it ought to have done. In order to account for this gradual diminution in the comet's period Encke conjectured the existence of a thin ethereal medium sufficiently dense to affect a light flimsy body like a comet, but incapable of obstructing a planet. It has been remarked by Hind that "this contraction of the orbit must be continually progressing, if we suppose the existence of such a medium; and we are naturally led to inquire, What will be the final consequence of this resistance? Though the catastrophe may be averted for many ages by the powerful attraction of the larger planets, especially Jupiter, will not the comet be at last precipitated on the Sun? The question is full of interest, though altogether open to conjecture."

Astronomers are not altogether agreed as to the propriety of this explanation. One argument against it is that with *perhaps* one exception none of the other short-period comets (all of them small and presumably deficient in density) seem affected as Encke's is. On the other hand Sir John Herschel favoured the explanation just given, as also does Hind who is the highest living authority on comets. A German mathematician, Von Asten, who devoted immense labour to the study of the orbit of Encke's Comet, thought there should be no hesitation in accepting the idea of a resisting medium, subject to the limitation that it does not extend beyond the orbit of Mercury. Von Asten's allusion to Mercury touches a subject which belongs more directly to the question of Mercury's orbit and to that other very interesting question, "Are there any planets, not at present known, revolving round the Sun within the orbit of Mercury."

Which is the largest and most magnificent comet recorded in history? It is virtually impossible to answer this question, because of the extravagant and inflated language made use of by ancient and medieval (I had almost added, and modern) writers. There is no doubt that the comet of 1680, studied by Sir I. Newton, the tail of which was curved, and from 70° to 90° long, must have been one of the finest on record, as it was also the one which came nearest to the Sun, for it almost grazed the Sun's surface.

The comet of 1744, visible as it was in broad daylight, was, no doubt, the finest comet of the 18th century, though in size it has been surpassed; yet its six tails must have made it a most remarkable object. So far as the 19th century is concerned, our choice lies between the comets of 1811, 1843, 1858, and 1861. The comet of 1811 is spoken of by Hind as "perhaps the most famous of modern times. Independently of its great magnitude, the position of the orbit and epoch of perihelion passage, were such as to render it a very splendid circumpolar object for some months." The tail as regards its length was not so very remarkable, for at its best, in October 1811, it was only about 25° long, its breadth, however, was very considerable; at one time 6°, the real length of the tail, about the middle of October, was more than 100,000,000 of miles, and its breadth about 15,000,000 of miles. The visibility of this comet was coincident with those events which proved to be the turning-point in the career of Napoleon I., and there were not wanting those who regarded the comet as a presage of his disastrous failure in Russia. Owing to the long period (17 months), during which this comet was visible,

FIG. 27.—The Great Comet of 1811.

it was possible to determine its orbit with unusual precision. Argelander found its period to be 3065 years, with no greater uncertainty than 43 years. The great dimensions of its orbit will be realised when it is stated that this comet recedes from the Sun to a distance of 14 times that of the planet Neptune.

Donati's comet of 1858, has already received a good deal of notice at my hands, but the question remains, what are its claims, to be regarded as the comet of the century, compared with that of 1843? It is not a little strange that though there must have been many persons who saw both, yet it was only quite recently that I came across, for the first time, a description of both these comets from the same pen. It ought, however, to be mentioned by way of explanation, that the inhabitants of Europe only saw the comet of 1843, when its brilliancy and the extent of its tail had materially diminished, about a fortnight after it was at its best.

The description of these two comets to which I have alluded, will be found in General J. A. Ewart's "*Story of a Soldier's Life*" published in 1881. Writing first of all of the comet of 1843, General Ewart says:—

"It was during our passage from the Cape of Good Hope to the Equator, and when not far from St. Helena, that we first came in sight of the great comet of 1843. In the first instance a small portion of the tail only was visible, at right angles to the horizon; but night after night as we sailed along, it gradually became larger and larger, till at last up came the head, or nucleus, as I ought properly to call it. It was a grand and wonderful sight, for the comet now extended the extraordinary

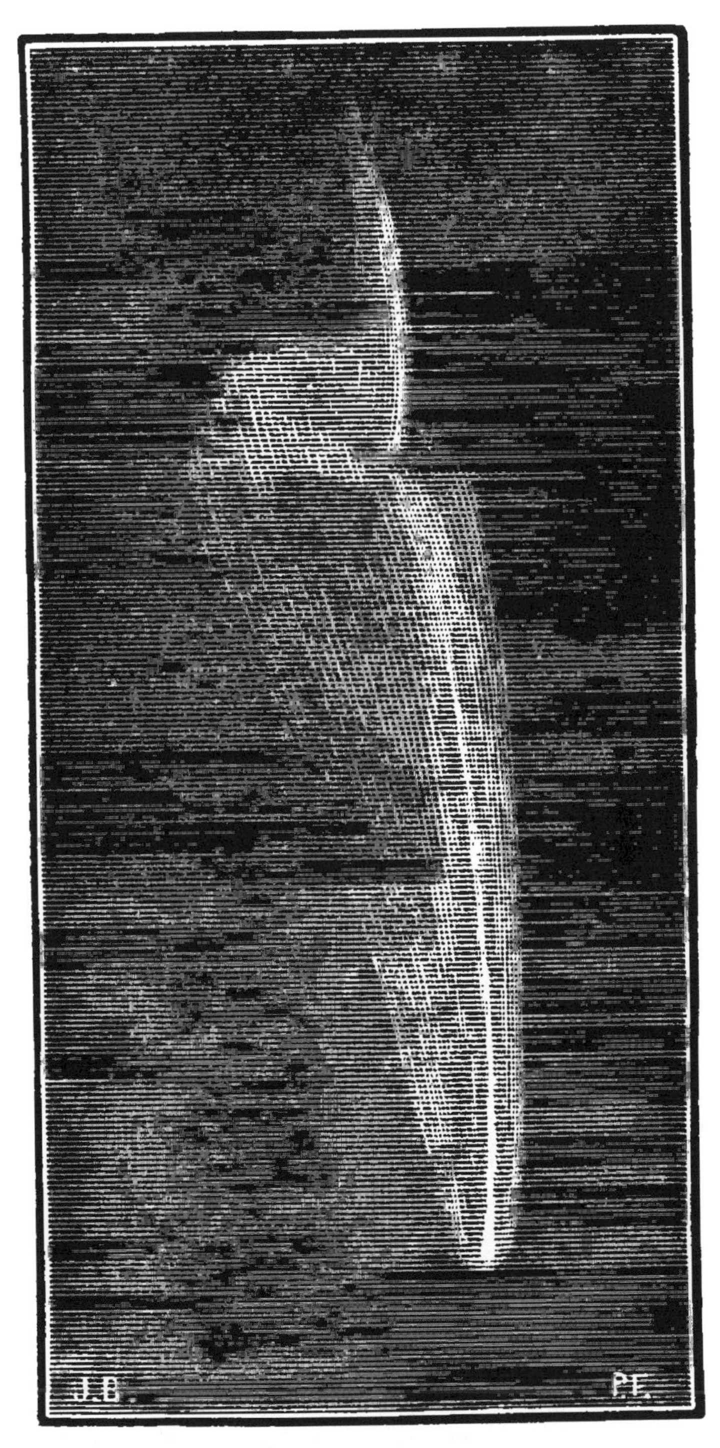

FIG. 28.—The Great Comet of 1882, on October 19 (Artus).

distance of one-third of the heavens, the nucleus being, perhaps, about the size of the planet Venus."—(Vol. i., p. 75.)

As regards Donati's comet of 1858, what the General says is:—

"A very large comet made its appearance about this time, and continued for several weeks to be a magnificent object at night; it was, however, *nothing to the one I had seen in the year* 1843, when on the other side of the equator."—(Vol. ii., p. 205.)

Passing over the great comet of 1861, on which I have already said a good deal, I must quit the subject of famous comets by a few words about that of 1882, which, though by no means one of the largest, was, in some respects, one of the most remarkable of modern times. It was visible for the long period of nine months, and was conspicuously prominent to the naked eye during September, but these facts, though note-worthy, would not have called for particular remark, had not the comet exhibited some special peculiarities which distinguished it from all others. In the first place, it seems to have undergone certain disruptive changes, in virtue of which the nucleus became split up into four independent nuclei. Then the tail may have been tubular, its extremtiy being not only bifid, but totally unsymmetrical with respect to the main part. The tubular character of the tail was suggested by Tempel. To other observers, this feature gave the idea of the comet, properly so-called, being enclosed in a cylindrical envelope, which completely surrounded the comet, and overlapped it for a considerable distance at both ends. Finally (and in this resembling Biela's comet) the comet of 1882 seems to have thrown off a fragment which became an independent body.

What has gone before, will, I think, have served abundantly to establish the position with which I started, namely, that comets occupy (and deservedly so) the front rank amongst those astronomical objects in which, on occasions, the general public takes an interest.

I have thus completed, so far as the space at my disposal has permitted, a popular descriptive Survey of the Solar System. Those who have perused the preceding pages, however slight may have been their previous acquaintance with the Science of Astronomy taken as a whole, will have no difficulty in realising that what I have said bears but a small proportion to what I have left unsaid. They will equally, I hope, be able to see, without indeed the necessity of a suggestion, that all those wondrous orbs which we call the planets could neither have come into existence nor have been maintained in their allotted places for so many thousands of years, except as the result of Design emanating from an All-powerful Creator.

Name.	Sidereal period.	Mean distance from Sun.	Diameter.	Surface. Earth = 1.	Volume. Earth = 1.	Mass. Earth = 1.	Density. Earth = 1.	Axial rotation.	Force of gravity. Fall: Ft. in 1 sec.	Velocity in orbit.
		Millions of miles.	Miles.					d. h. m.		Miles per hour
SUN	......	....	866,200	11,946	1,305,000	332,000	0.25	25 7 48	461	
								h. m. s.		
MERCURY	88 days.	36	3,008	0.144	0.055	0.066	1.26	24 5 30	7	107,000
VENUS	225 "	67	7,480	0.891	0.841	0.782	0.92	23 21 23	14	78,000
EARTH	365 "	93	7,926	1.000	1.000	1.000	1.00	23 56 4	16	66,000
MARS	687 "	141	5,000	0.398	0.251	0.107	0.45	24 37 23	4	53,000
Minor Planets. EROS (433)	1.76 years.	....	18	Eros is the nearest to the Sun of the Minor Planets, part of its orbit falling between the Earth and Mars.						
Minor Planets. VESTA (4)	3.6 "	219	214	Vesta is the largest of the Minor Planets.						
Minor Planets. THULE (279)	8.8 "	396	....	Thule is the most distant from the Sun of the Minor Pl[illegible]						

NAME.	Sidereal period.	Distance from Earth.	Diameter.	Surface. Earth=1.	Volume. Earth=1.	Mass. Earth=1.	Density, Earth=1.	Axial Rotation.	Force of gravity. Fall: Feet in 1 sec.	Velocity in orbit.
	d. h. m.	Miles.	Miles.					d. h. m.		Miles per hour.
MOON........	27 7 43	237,300	2,160	0.074	0.02034	0.0128	0.63	27 7 43	2.48	2,273

SATELLITES OF MARS.

NAME.	Discoverer.	Mean distance from Mars.	Sidereal period.	Diameter.	Maximum elongation.	Apparent star magnitude.
		Miles.	d. h. m.	Miles.	″	
1. PHOBOS.....	A. Hall, August 17, 1877 .	6,000	0 7 39	7	12	11½
2. DEIMOS.....	A. Hall, August 11, 1877 ..	15,000	1 6 18	6	32	13½

SATELLITES OF JUPITER.

NAME.	Discoverer.	Mean distance from Jupiter.	Sidereal period.	Diameter.	Apparent diameter of Jupiter seen from satellite.	Apparent star magnitude.
		Miles.	d. h. m.	Miles.	° ″	
5.	Barnard	....	0 11 49	?	......	..
1. IO............	....	267,000	1 18 29	2,390	19 49	7
2. EUROPA	Galileo, January 7–13, 1610.	425,000	3 13 13	2,120	12 25	7
3. GANYMEDE	....	678,000	7 4 0	3,980	7 47	6
4. CALLISTO ..	....	1,192,000	16 18 5	2,970	4 25	7

NAME.	Discoverer.		Mean distance from Saturn.	Sidereal period.			Diameter.	Apparent diameter of Saturn seen from satellite.	Apparent star magnitude.
			Miles.	d.	h.	m.	Miles.	°	
1. MIMAS.......	Sir W. Herschel,	Sept, 17, 1789	115,000	0	22	37	1,000	33	17
2. ENCELADUS	" "	Aug. 28, 1789	147,000	1	8	53	?	26	15
3. TETHYS.....	J. D. Cassini,	March, 1684	183,000	1	21	18	500	21	13
4. DIONE.......	" "	March, 1684	234,000	2	17	41	500	16	12
5. RHEA........	" "	Dec. 23, 1672	327,000	4	12	25	1,200	12	10
6. TITAN.......	Huygens,	Mar. 25, 1655	758,000	15	22	41	3,300	5	8
7. HYPERION..	W. Bond & Lassell,	Sept. 19, 1848	916,000	21	7	7	?	4	17
8. IAPETUS....	J. D. Cassini,	Oct. 25, 1671	2,221,000	79	7	53	1,800	2	9

SATELLITES OF URANUS.

NAME.	Discoverer.	Mean distance from Uranus.	Sidereal period.			Maximum elongation
		Miles.	d.	h.	m.	"
1. ARIEL................	Lassell, September 14, 1847............	124,000	2	12	28	12
2. UMBRIEL............	O. Struve, October 8, 1847............	173,000	4	3	27	15
3. TITANIA.	Sir W. Herschel, January 11, 1787.....	285,000	8	16	55	33
4. OBERON	" "	381,000	13	11	6	44

GENERAL INDEX.

V.

W.

Y.

Z.

(5)

THE END.